ARAB OIL
and
UNITED STATES
ENERGY REQUIREMENTS

Abbas Alnasrawi

ASSOCIATION OF ARAB-AMERICAN UNIVERSITY GRADUATES, INC.

Belmont, Massachusetts, 1982

First published in the United States of America in August, 1982 by
The Association of Arab-American University Graduates, Inc.

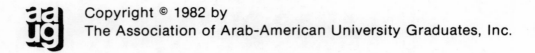
ISBN 0-937694-52-5

Abbas Alnasrawi is Professor of Economics at the University of Vermont, Burlington, Vermont, U.S.A.

Design and typesetting by Accugraphics

2219398

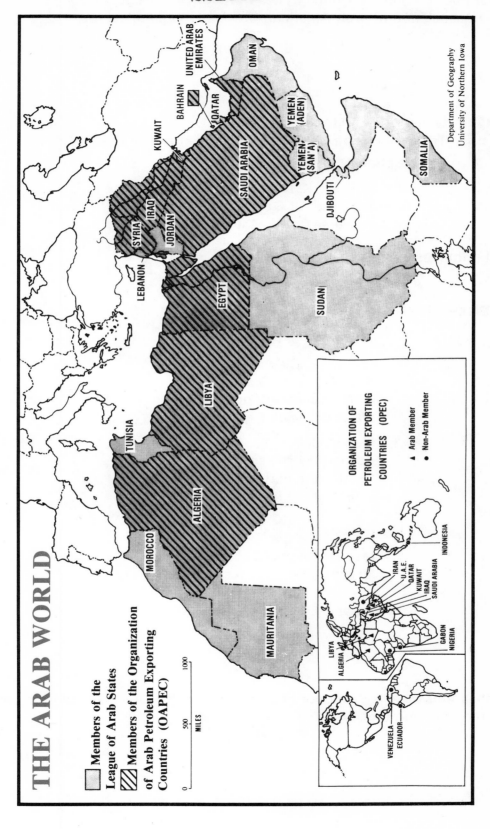

THE ARAB WORLD

Members of the
League of Arab States

Members of the Organization
of Arab Petroleum Exporting
Countries (OAPEC)

MILES

0   500   1000

ORGANIZATION OF
PETROLEUM EXPORTING
COUNTRIES (OPEC)

▲ Arab Member
● Non-Arab Member

Department of Geography
University of Northern Iowa

Editorial Supervision

Basheer K. Nijim
Professor of Geography
University of Northern Iowa

Editorial Assistance

Jeffrey J. Blaga, Chairman
Social Studies Area
Price Laboratory School
University of Northern Iowa

William E. Blake
Social Studies Consultant
Area Education Agency 7
Waterloo, Iowa District

Marvin Heller
Associate Professor of
Curriculum and Instruction
University of Northern Iowa

Illustrations

Map: Sherijo Dullard
Graphs: Bonnie Sines and Sherijo Dullard

Typing and Editorial Assistance

Sandra Heller

INTRODUCTION

It is widely believed that the "energy crisis" of the 1970s resulted from the embargo measures carried out by Arab oil producing countries following the October 1973 war between Israel on the one hand and Egypt and Syria on the other, and that the "crisis" was made worse by subsequent increases in oil prices. A careful reading of the energy history in general and that of oil in particular in the twentieth century reveals that an energy imbalance had been in the making for decades. The imbalance was bound to happen because of the phenomenal growth after World War Two in the demand for energy in general but especially for oil. The increase in demand for energy became primarily a demand for oil. This demand was needed for the reconstruction of the European and Japanese economies, for their continued economic growth, and for the continued economic growth occurring in the United States and in the rest of the world.

One of the most striking developments since World War Two was the change in the contribution of various forms of energy to total consumption of energy. The prosperous and vitally important coal industry decreased in importance relative to more versatile, more convenient, and cheaper forms of energy, especially oil and natural gas. Consequently, Western economies became increasingly dependent on oil, whose major reserves were located in other parts of the world. The fact that major centers of energy consumption were quite distant from the sources of oil supply was a major cause of the increasing concerns about oil availability and price.

In this study we shall examine a number of factors that are involved in an economic interdependence between the United States as the major oil importing country in the world and the Arab oil producing region as the major supplier of oil.

Part I deals with the subject of energy and oil in general. It will deal with total consumption of energy, various forms of energy, energy's relationship to national output, the contribution of various energy forms to consumption between 1950 and 1979, and the situation of the major consuming and producing areas of the world. Part II focuses on issues related to the United States. It deals with United States' consumption, production, and import of energy. It also considers American-Arab trade relations in terms of both energy and other goods and services. Part III highlights the role of oil in economic and social development in Arab oil producing countries. Special attention is given to the role of oil revenues as the major source of foreign exchange earnings as well as the major source of finance for current budgetary programs and development plans. In the last ten years, however, oil revenues have assumed a new role in economies outside the political boundaries of the Arab countries. This new role found its expression in the flows of financial aid from the oil producing countries to other developing countries; in the expansion of trade with these countries as well as through the transfer of funds by nationals of non oil producing countries who were attracted to employment opportunities in the Arab oil producing countries.

Table of Contents

List of Figures

List of Tables

PART I

ENERGY: IMPORTANCE, SOURCES AND CONSUMPTION PATTERNS

GENERAL OVERVIEW

A study by the United States Senate Committee on Energy and Natural Resources made the important statement that without oil no modern economy could exist. The study concluded that at its most fundamental level the energy problem was a problem of oil. These observations were based on the fact that oil had become not only the most important source of energy throughout the world but that it had also become an indispensable raw material for a large number of products in the modern economy. To appreciate the importance of oil one only needs to look at the vast assortment of petrochemical-based products, from textiles to medicines and from fertilizers to precision instruments. An obvious question is why oil has assumed such critical importance. The answer lies in the fact that over the decades, but especially in this century, oil was found to have superior qualities over other forms of energy.

It goes without saying that world economic growth has been intimately associated with the development of the energy sector, primarily coal. Since the era of industrial revolution coal has played a critical role first in the development of the British economy and later in other economies. With the start of oil production, first in the United States in 1860 and later in other countries, the make-up of energy consumption for industrial and non-industrial uses began to change. The coming of the automobile era and the movement toward mechanization in all economies only increased the demand for oil as the only fuel that could meet this kind of demand. World demand for energy, especially oil, increased greatly in the years following World War Two.

Most of this increase took place in the industrialized countries (see Tables 1 and 2, and Figures 1). In the United States the consumption of energy between 1950 and 1960 went from 16.2 million barrels per day of oil equivalent (MBDOE) to 21.8 MBDOE, a 35 percent increase. During the same period consumption of oil alone rose from 6.4 MBD to 9.7 MBD, a 52 percent increase. In Japan, the increase in energy consumption (and particularly of oil) was nothing but phenomenal: it rose from .7 MBDOE to 1.7 MBDOE, or an increase of 143 percent. Even more dramatic was the increase of Japan's oil consumption from less than 0.1 MBD to .64 MBD. In Western Europe the 54 percent increase in energy consumption was less than that of Japan but more than that of the United States: from 7.2 MBDOE in 1950 to 11.1 MBDOE in 1960. Oil consumption however increased from 1.2 MBD in 1950 to 4.1 MBD in 1960, or by 242 percent.

It is clear from these data and from the fact that oil consumption could not be met from domestic sources that a pattern of economic interdependence had to develop between oil producing countries on the one hand and oil consuming countries on the other. Given the fact that enormous reserves of oil are found in certain parts of the Arab world, Arab oil exports acquired a rapidly increasing importance. Before dealing with specifics of oil interdependence between the United States and the Arab World, it will be useful to consider changes in the relative importance of various forms of energy during the 1960s and 1970s. A brief survey of the relationship between energy consumption and the gross national product will also be made.

4

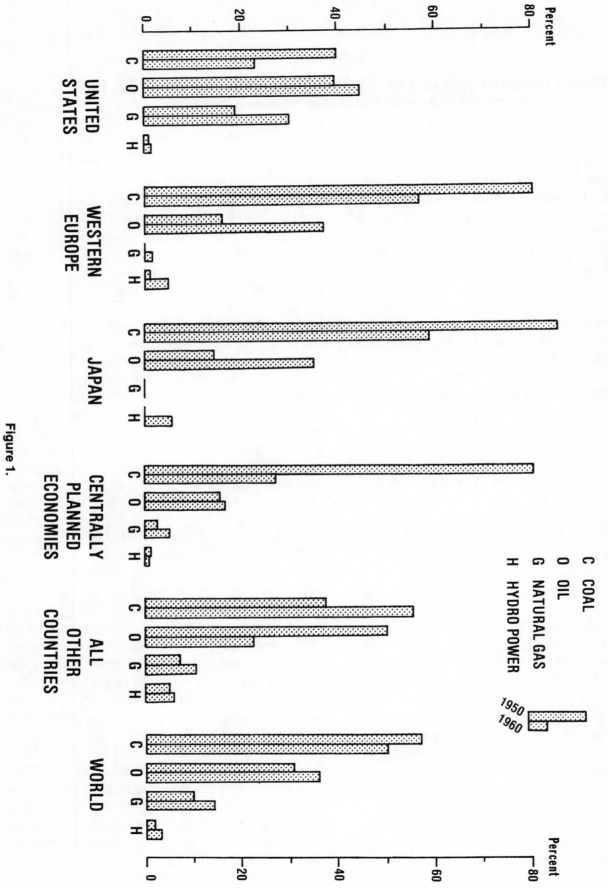

WORLD CONSUMPTION OF ENERGY BY SOURCE AND MAJOR REGION, 1950 and 1960.

Figure 1.

Table 1.

WORLD CONSUMPTION OF ENERGY BY SOURCE AND MAJOR REGION, 1950
(in millions of barrels per day of oil equivalent, or MBDOE)

	Coal	Oil	Natural Gas	Hydro Power	Total
United States	6.5	6.4	3.1	.2	16.2
Western Europe	5.8	1.2	N	.2	7.2
Japan	.6	.1	N	N	.7
Subtotal	12.9	7.7	3.1	.4	24.1
Centrally Planned Economies	6.1	1.2	.2	.1	7.6
Rest of the World	1.5	2.0	.3	.2	4.0
World Total	20.5	10.9	3.6	.7	35.7

N = Negligible

Source: Derived from OECD, *Energy Policy: Problems and Objectives*

Table 2.

WORLD CONSUMPTION OF ENERGY BY SOURCE AND MAJOR REGION, 1960
(in million barrels per day of oil equivalent, or MBDOE)

	Coal	Oil	Natural Gas	Hydro Power	Total
United States	5.1	9.7	6.6	.4	21.8
Western Europe	6.3	4.1	.2	.5	11.1
Japan	1.0	.6	---	.1	1.7
Subtotal	12.4	14.4	6.8	1.0	34.6
Centrally Planned Economies	14.2	3.0	1.0	.2	18.4
Rest of the World	3.8	4.1	.7	.4	6.8
World Total	30.4	21.5	8.5	1.6	59.7

Source: See source for Table 1.

Energy Consumption

Following World War II there was a steady increase in the demand for energy. In 1973 consumption patterns began to change, and that year can be regarded as a turning point in the trend of energy demand. Thus it will be useful to distinguish between the periods before and after 1973. Specifically, a comparison will be made between the periods from 1965 to 1973 and from 1973 to 1979. During the earlier period world energy consumption increased at an annual rate of 6.1 percent. Following the oil price revolution of 1973, the rate decreased to about 3.1 percent per year.

Growth in the demand for each major source of energy is given in Table 3.

Oil and coal provided most of the world energy supplies during the 1960s. Oil production in 1965 was about 31 million barrels per day, accounting for approximately 39 percent of total world energy consumption. Coal also provided 38 percent of world energy consumption or about 31 MBDOE. Together these two sources supplied 77 percent of world energy consumption at that time. As the demand for energy continued to increase it was primarily oil that was called

Table 3.

WORLD CONSUMPTION OF ENERGY BY SOURCE, 1965–1979
(in millions of barrels per day of oil equivalent, or MBDOE)

	1965	1970	1973	1975	1977	1979	Average Rate of 1965–73	Annual Change 1973–79
Oil	31.1	46.3	56.5	55.4	61.2	64.1	10.2	2.3
Natural Gas	13.0	19.2	21.6	22.3	23.9	26.1	8.3	3.5
Coal	31.0	33.0	33.6	34.5	36.9	39.7	1.1	3.0
Hydro Power	4.7	6.1	6.6	7.1	7.5	8.3	5.1	4.3
Nuclear Energy	.3	.4	1.0	1.7	2.7	3.1	29.2	35.0
Total	80.1	105.0	119.3	121.0	132.2	141.3	6.1	3.1

Notes on Sources for Tables 3–18:
Throughout Part I two publications were used as the main sources for the data. Those publications are:

1. British Petroleum, *BP Statistical Review of the World Oil Industry.* This review is published annually.

2. U.S. Department of Energy, *Annual Report to Congress, 1979.*

It was necessary to convert data that were reported in different units (ton of oil equivalent or ton of coal equivalent) to one common unit, i.e., MBDOE or millions of barrels per day of oil equivalent. It should be noted that due to this conversion and to rounding, variations in the data may be found.

upon to meet the rising demand. By 1973 oil consumption accounted for approximately 47.4 percent of world total energy consumption. Table 4 provides data on the comparative contribution of various forms of energy, and Figure 2 gives a visual portrayal of the trends. The 1970s show a continued reliance on oil and natural gas to provide for the increases in the world energy demand. By 1979 oil and natural gas together provided approximately 64 percent of world energy consumption (85 MBDOE) as compared to 55 percent (44 MBDOE) in 1965.

Following this summary of major changes in the total energy picture, changes in the demand for individual sources of energy will now be considered.

Oil

Oil maintained its position as the major source of energy consumption throughout the period 1960 to 1979. Its share of world energy consumption rose from about 35 percent in 1960 to over 45 percent in 1979. Note, however, that after 1973, when oil accounted for more than 47 percent of total consumption, there was a slight reduction in its share. As given in Table

3, it can be seen from data between 1965 and 1973 oil consumption increased at an annual rate of 10.2 percent, a rate much higher than the 6.1 percent per year for total energy growth. During 1973–1979, however, these trends changed when oil consumption grew at an annual rate of 2.3 percent only compared to an overall growth of consumption of 3 percent per annum. Thus, the data for the second half of the 1970s seem to indicate a change in the patterns of world energy consumption.

Geographically, there is a great variation in oil consumption patterns (Tables 5 and 6, and Figure 3). The greatest single consumer was and continued to be the United States. The United States accounted for 45 percent of the world total oil demand in 1960 and, while this figure was reduced to 30 percent in 1973 and 28 percent in 1979, the United States was still the major oil consuming country in the world. Unlike the United States, Western Europe experienced a growth in its share of world oil consumption during this period. It consumed almost 19 percent of world oil supplies in 1960, and this proportion rose to over 23 percent by 1979. Japan also increased its share of world oil consumption during the 1960s and 1970s. From a level of approximately 3 percent in 1960, its consumption increased to about 9 percent in

Table 4.

WORLD ENERGY CONSUMPTION: PERCENTAGE CONTRIBUTION BY SOURCE, 1965–1979

	1965	1970	1973	1975	1977	1979
Oil	38.8	44.1	47.4	45.8	46.3	45.4
Natural Gas	16.2	18.3	18.1	18.4	18.1	18.5
Coal	38.4	31.0	28.2	28.5	28.0	28.1
Hydro Power	5.9	5.8	5.5	5.9	5.7	5.9
Nuclear Energy	.4	.4	.8	1.4	2.0	2.2

Source: Computed from Table 3.

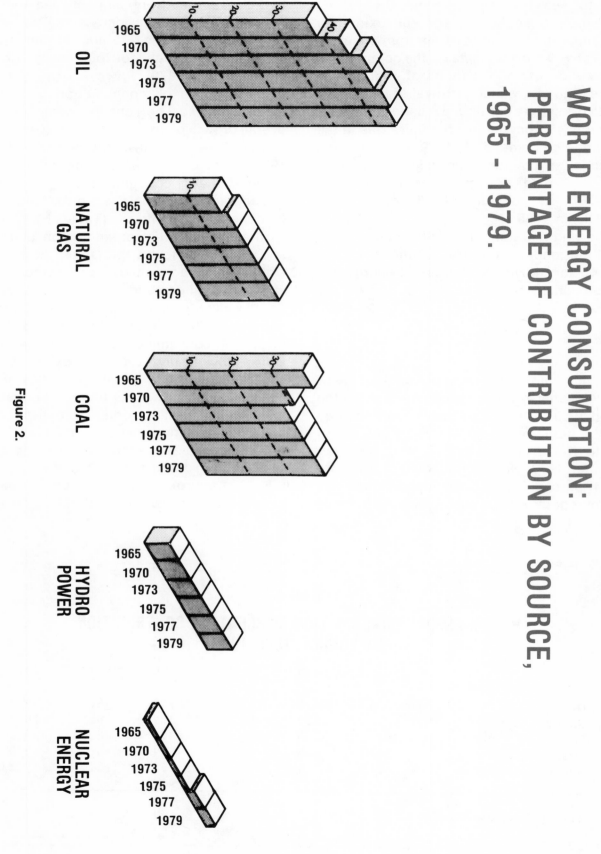

WORLD ENERGY CONSUMPTION:
PERCENTAGE OF CONTRIBUTION BY SOURCE,
1965 - 1979.

Figure 2.

Table 5.

OIL: WORLD CONSUMPTION BY MAJOR REGION, 1965–1979
(in million barrels per day, or MBD)

	1965	1973	1979	Percent Average Annual Rate of Change 1965–73	1973–79
United States	11.3	16.9	17.9	6.2	1.0
Western Europe	7.8	15.2	14.9	11.9	- .3
Japan	1.8	5.5	5.5	25.7	0
Subtotal	20.9	37.6	38.3	10.0	.3
Centrally Planned Economies	4.5	9.2	12.8	13.1	6.5
Rest of the World	5.7	10.2	13.0	3.2	0
World Total	31.1	57.0	64.1	8.8	2.1

Table 6.

WORLD OIL CONSUMPTION: PERCENTAGE DISTRIBUTION BY MAJOR REGION, 1965–1979

	1965	1973	1979
United States	33.7	29.6	27.9
Western Europe	23.4	26.7	23.2
Japan	5.4	9.6	8.9
Subtotal	62.5	66.0	59.8
Centrally Planned Economies	13.4	16.1	20.0
Rest of the World	24.2	18.0	20.1
World Total	100	100	100

Source: Computed from Table 5.

WORLD OIL CONSUMPTION: PERCENTAGE DISTRIBUTION BY MAJOR REGION, 1965, 1973, and 1979

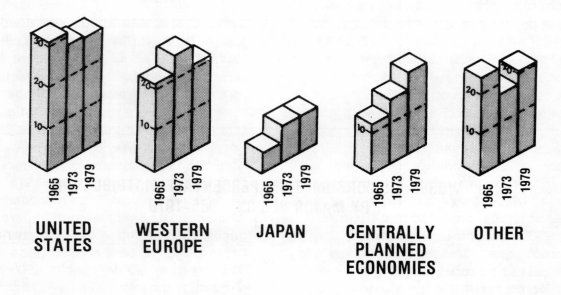

Figure 3.

1979. In absolute terms, Japan's oil consumption increased from 1.8 million barrels per day in 1965 to 5.4 MBD in 1979.

The Centrally Planned Economies (CPE) experienced a growth in their share of world oil consumption during this period, from about 13 percent in 1965 to 20 percent in 1979, or from 4.5 MBD in 1965 to 12.8 MBD in 1979.

Natural Gas

World consumption of natural gas increased during the period 1975-1979 at a higher rate than that of total energy consumption. These different rates of growth allowed natural gas to increase its share of total energy consumption from 16.2 percent in 1965 to 18.5 percent in 1979 (Table 4). The United States remained the largest single consumer of natural gas throughout the 1960s and 1970s (Tables 7 and 8, and Figure 4). In 1979 it accounted for over 38 percent of total world consumption. Natural gas was also an important contributor to Western Europe's energy consumption. Its share in that region's consumption of energy increased from 3 percent in 1965 to 14 percent in 1979. It should be noted that between 1965 and 1973 consumption of natural gas in Western Europe increased at an annual rate of 69 percent compared to a world average growth rate of only 8 percent per year. From 1973 to 1979 natural gas consumption in Western Europe increased at an annual rate of 7.1 percent but was still above the 3.4 percent rate of increase for the world as a whole. Japan's consumption of natural gas also increased during 1965-1979. In 1965, Japan accounted for only 0.8 percent of world consumption, and by 1979 this share had increased to 1.5 percent. As to the Centrally Planned Economies, their consumption of natural gas also increased. The annual rate of growth in consumption was about 10.3 percent during the period 1965 to 1973 and 11.8 percent for 1973-1979. Their share of world natural gas consumption increased from 21 percent in 1965 to 33 percent in 1979.

Coal

Coal continues to supply a large share of world energy demand. In 1979 it accounted for 28 percent of world energy consumption (Table 4 and Figure 2). World coal production increased from approximately 32 MBDOE in 1965 to almost 40 MBDOE by the year 1979. In terms of demand, the annual rate of increase was 1 percent during 1965-1973 and rose to 3 percent per year for the period 1973-1979.

United States' coal consumption increased from almost 6.2 MBDOE in 1979 at a rate of 1 percent per year from 1965 to 1973 and about 2 percent annually from 1973 to 1979 (Tables 9 and 10, and Figure 5). United States share of world coal consumption remained relatively stable throughout the period, rising only slightly from about 19.2 percent in 1965 to about 19.4 percent in 1979. In Western Europe on the other hand the share of coal consumption to total energy consumption declined rapidly in the 1960s and the 1970s. In 1965 coal provided 60 percent of total energy consumption. By 1973 that share had declined to 30 percent, and to 20 percent by 1979. Relative to world consumption, coal consumption in Western Europe represented 21 percent in 1965 but only 13 percent in 1979. Japan increased its share of world coal consumption by a small fraction from 2.8 percent in 1965 to about 3 percent by the year 1979. By 1979 coal supplied about 15 percent of Japan's total energy consumption compared with 29 percent in 1965. The Centrally Planned Economies on the other hand consumed the greatest share of world coal supplies during 1965-1979: approximately 44 percent of world coal consumption in 1965 and 51 percent in 1979.

Nuclear Energy

In its early years, nuclear energy was described as the panacea for the industrialized world, and during the period 1965 to 1979 it enjoyed a phenomenal rate of growth. It maintained the highest growth rate among

Table 7.

NATURAL GAS: WORLD CONSUMPTION BY MAJOR REGION, 1965–1979
(in millions of barrels per day of oil equivalent, or MBDOE)

	1965	1973	1979	Percent Average Annual Rate of Change 1965–73	1973–79
United States	8.7	11.5	10.0	4.0	- 2.2
Western Europe	.4	2.6	3.7	69.0	7.1
Japan	.1	.1	.4	0	21.4
Subtotal	9.2	14.2	14.1	6.7	.2
Centrally Planned Economies	2.8	5.1	8.7	10.3	11.8
Rest of the World	1.2	2.4	3.3	12.5	6.2
World Total	13.2	21.7	26.1	8.0	3.4

Table 8.

WORLD NATURAL GAS CONSUMPTION: PERCENTAGE DISTRIBUTION BY MAJOR REGION, 1965–1979

	1965	1973	1979
United States	65.9	53.0	38.3
Western Europe	3.0	12.0	14.2
Japan	.8	.4	1.5
Subtotal	69.7	65.4	54.0
Centrally Planned Economies	21.2	23.5	33.3
Rest of the World	9.1	11.1	12.6
World Total	100	100	100

Source: Computed from Table 7.

WORLD NATURAL GAS CONSUMPTION: PERCENTAGE DISTRIBUTION BY MAJOR REGION, 1965, 1973, and 1979

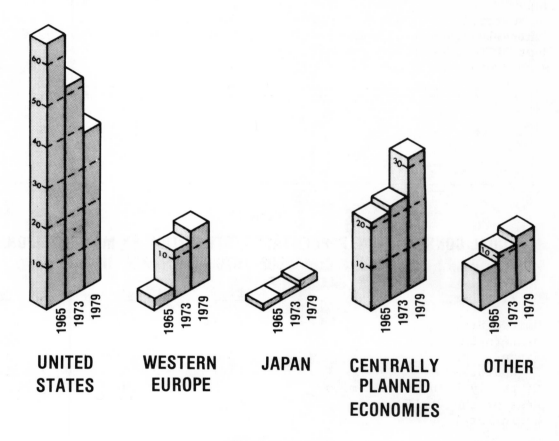

Figure 4.

Table 9.

COAL: WORLD CONSUMPTION BY MAJOR REGION, 1965–1979
(in millions of barrels per day of oil equivalent, or MBDOE)

	1965	1973	1979	Percent Average Annual Rate of Change	
				1965–73	1973–79
United States	6.1	6.7	7.7	1.2	2.4
Western Europe	6.7	5.1	5.3	−3.0	.1
Japan	.9	1.2	1.2	3.8	0
Subtotal	13.7	13.0	14.2	− .1	1.5
Centrally Planned Economies	14.1	16.9	20.6	2.5	3.6
Rest of the World	3.9	3.7	4.9	− .1	5.4
World Total	31.7	33.6	39.7	.1	3.0

Table 10.

COAL CONSUMPTION: PERCENTAGE DISTRIBUTION BY MAJOR REGION, 1965–1979

	1965	1973	1979
United States	19.2	19.9	19.4
Western Europe	21.1	15.2	13.4
Japan	2.8	3.6	3.0
Subtotal	43.9	38.6	35.8
Centrally Planned Economies	44.5	50.3	51.9
Rest of the World	12.3	11.1	12.3
World Total	100	100	100

Source: Computed from Table 9.

WORLD COAL CONSUMPTION: PERCENTAGE DISTRIBUTION BY MAJOR REGION, 1965, 1973, and 1979

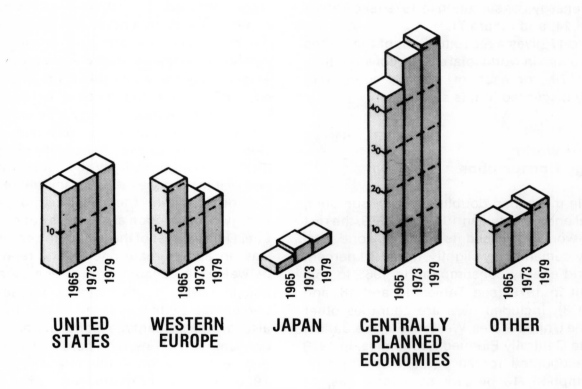

Figure 5.

energy sources, rising by 35 percent per year during 1973–1979 (Table 3). Despite this tremendous growth nuclear energy accounted for only 3.1 percent of world consumption of energy (see Tables 11 and 12, and Figure 6).

Hydro Power

This important source of energy grew at an annual rate of 5 percent during the period 1965–1973 and of 4.3 percent between 1973 and 1979. It contributed about 6 percent to total world energy consumption in 1979 (see Tables 13 and 14, and Figure 7).

Table 17 gives a recapitulation of the United States share in world total energy consumption, 1965–1979, for each of the five sources of energy discussed in this study.

Rest of World Energy Consumption

While more than doubling the consumption of total energy between 1965 and 1979, the rest of the world increased its share of world total energy demand only slightly, from 12.1 percent of world energy consumption in 1965 to 15.5 percent in 1979 (see Tables 15 and 16, and Figure 8). Included here are countries other than the United States, Western Europe, Japan, and the Centrally Planned Economies. In 1979 they accounted for 20 percent of world oil consumption, 13 percent of natural gas, 12 percent of coal, and 35 percent of hydro power.

Energy-Gross National Product Relationship

The gross national product (GNP) is defined as the market value (i.e., the price paid by the final consumer) of all the goods and services produced by a country's economy in any one year. In the process of producing the national output (or GNP) a certain amount of energy is consumed. For instance, in 1973 the United States economy needed close to the energy equivalent of 37 million barrels per day of oil to produce a national output the market value of which was $1,326 billion (or $1.3 trillion). Since energy by definition is the power that is needed to run the economy it is clear that there is a direct link between the level of national output and the energy that is required to produce that output. This relationship is expressed by the Energy/GNP ratio. Clearly, therefore, for any amount of output it would be beneficial to the economy if the amount of energy consumed were reduced. In other words the more expensive a given amount of energy (say one barrel of oil) the stronger would be the incentive for the economy to be more energy efficient. By the same token the cheaper the cost of energy the weaker would be the incentive to be more energy efficient, i.e., the stronger the inclination to waste energy. Consequently, as the cost of energy goes up attempts will be made by the economy to use less of it to produce the same amount of output. Obviously, then, the relationship between energy consumption and GNP is not a constant one. During most of the twentieth century there was an approximate one to one relationship between the consumption of energy and the rise in GNP. In other words, for each one percentage point increase in GNP there was also a one percentage point increase in the consumption of energy. There was, however, a degree of variation in this relationship. Up to 1920, energy consumption growth was approximately 1.5 times greater than GNP growth due primarily to a rapid growth in the manufacturing sector. The period between 1920 and 1965 marked a decline in the energy growth to approximately 0.75 times GNP growth due to increased fuel efficiency. After 1965, the Energy/GNP relationship increased, with energy growth reaching 1.4 times the growth in GNP. This change reflected losses in electricity conversions and rising petroleum consumption in the transportation sector. The events of the oil price shock in 1973 and 1974 were instrumental in causing a significant

Table 11.

NUCLEAR ENERGY: WORLD CONSUMPTION BY MAJOR REGION, 1965–1979
(in millions of barrels per day of oil equivalent, or MBDOE)

	1965	1973	1979	Percent Average Annual Rate of Change 1965–73	Percent Average Annual Rate of Change 1973–79
United States	.1	.4	1.5	37.5	45.8
Western Europe	.2	.3	.8	6.3	27.8
Japan	---	---	.3	---	---
Subtotal	.3	.7	2.6	16.7	45.2
Centrally Planned Economies	---	---	.3	---	---
Rest of the World	---	.3	.2	---	5.6
World Total	.3	1.0	3.1	29.2	35.0

Table 12.

WORLD NUCLEAR ENERGY CONSUMPTION: PERCENTAGE DISTRIBUTION BY MAJOR REGION, 1965–1979

	1965	1973	1979
United States	33.3	40.0	48.4
Western Europe	66.7	30.0	25.8
Japan	---	---	9.7
Subtotal	100	70.0	83.9
Centrally Planned Economies	---	---	9.7
Rest of the World	---	30.0	6.4
World Total	100	100	100

Source: Computed from Table 11.

WORLD NUCLEAR ENERGY CONSUMPTION: PERCENTAGE DISTRIBUTION BY MAJOR REGION, 1965, 1973, and 1979

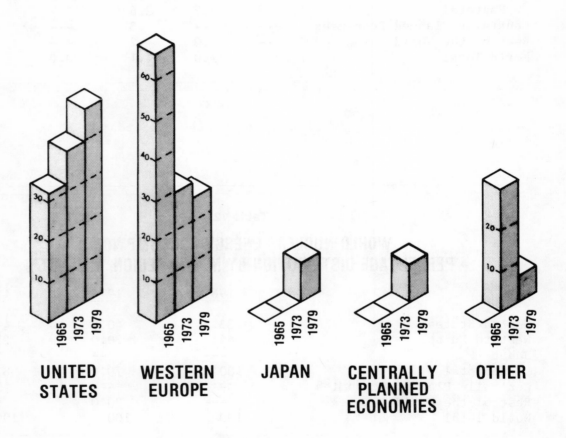

Figure 6.

Table 13.

HYDRO POWER: WORLD CONSUMPTION BY MAJOR REGION, 1965–1979
(in millions of barrels per day of oil equivalent, or MBDOE)

	1965	1973	1979	Percent Average Annual Rate of Change 1965–73	1973–79
United States	1.0	1.5	1.6	6.3	.8
Western Europe	1.5	1.9	2.2	3.3	2.6
Japan	.4	.3	.4	-3.1	5.5
Subtotal	2.9	3.7	4.2	3.4	2.3
Centrally Planned Economies	.6	.9	1.2	6.3	5.6
Rest of the World	1.2	2.0	2.9	8.3	7.5
World Total	4.7	6.6	8.3	5.1	4.3

Table 14.

WORLD CONSUMPTION OF HYDRO POWER:
PERCENTAGE DISTRIBUTION BY MAJOR REGION, 1965–1979

	1965	1973	1979
United States	21.2	22.7	19.3
Western Europe	31.9	28.9	26.5
Japan	8.5	4.5	5.0
Subtotal	61.7	56.1	50.6
Centrally Planned Economies	12.8	13.6	14.5
Rest of the World	25.5	30.3	34.9
World Total	100	100	100

Source: Computed from Table 13.

WORLD CONSUMPTION OF HYDRO POWER: PERCENTAGE DISTRIBUTION BY MAJOR REGION, 1965, 1973, and 1979

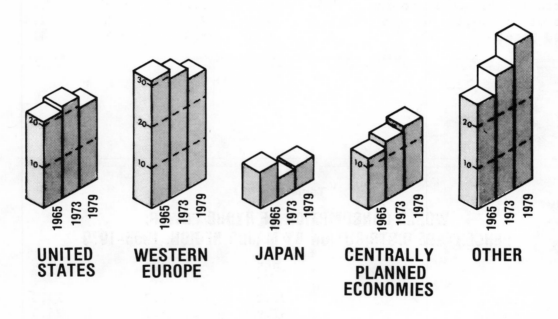

Figure 7.

Table 15.

WORLD ENERGY CONSUMPTION BY MAJOR REGION, 1965–1979
(in millions of barrels per day of oil equivalent, or MBDOE)

	1965	1973	1979	Percent Average Annual Rate of Change 1965–73	1973–79
United States	26.9	36.6	38.1	4.5	.1
Western Europe	17.1	25.0	26.6	5.8	.1
Japan	3.1	7.1	7.7	16.1	1.4
Subtotal	47.1	68.7	72.4	5.7	.9
Centrally Planned Economies	23.3	32.1	43.6	4.7	6.0
Rest of the World	9.7	18.5	25.3	11.3	6.1
World Total	80.1	119.3	141.3	6.1	3.1

Table 16.

WORLD ENERGY CONSUMPTION: PERCENTAGE DISTRIBUTION BY MAJOR REGION, 1965–1979

	1965	1973	1979
United States	33.6	30.7	27.0
Western Europe	21.3	21.0	18.8
Japan	3.9	6.0	5.4
Subtotal	58.8	57.7	51.2
Centrally Planned Economies	29.0	26.9	30.9
Rest of the World	12.1	15.5	17.9
World Total	100	100	100

Source: Derived from data in Table 3 and Table 15.

WORLD ENERGY CONSUMPTION: PERCENTAGE DISTRIBUTION BY MAJOR REGION, 1965, 1973, and 1979.

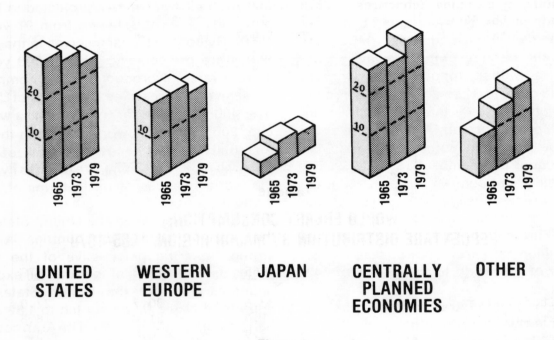

Figure 8.

reduction in energy use by 1980 relative to historical patterns.

In sum, it can be said that, in the long-run, Energy/GNP relationships are not constant due to changes in technology, productivity, conservation measures, and changes in the mix of goods and services produced.

Arab Oil in the Context of World Energy

In 1979 the Arab oil producing countries had a combined crude oil output of 22 million barrels per day (MBD). This amount represented about 16 percent of the total world consumption of energy, which was 140 million barrels per day of oil equivalent (MBDOE). Relative to the world oil output, however, Arab oil assumes higher importance in that it represented close to 34 percent of the world oil output which reached a total of 65.7 MBD in that year. More importantly Arab oil derives its significance from the fact that the Arab oil exporting countries, members of the Organization of Arab Petroleum Exporting Countries (OAPEC),* are responsible for 63 percent of the crude oil that moves in international trade channels. Furthermore, these Arab countries have 338 billion barrels of crude oil reserves, or 53 percent of the world total reserves of 642 billion barrels.

In addition to oil the Arab countries possess considerable reserves of natural gas amounting to 16 percent of the world total. In 1979 the Arab region produced 11 percent of the world output of natural gas but was able to consume only 6 percent of the world total consumption.

It is important to note that the Arab region's consumption of oil is only a small fraction of the world total consumption. Thus in 1979 while contributing over one third of the world output of oil the Arab region's consumption of oil was only 2 percent of the world total. Western Europe on the other hand was responsible in that year for 23 percent of the world consumption of oil while contributing less than 4 percent to the world output of oil.

The disparity between the rates of consumption and production of both oil and natural gas in the Arab region means, among other things, a fast rate of depletion of these natural resources to meet the world demand for energy. Current rates of production indicate that Arab oil reserves are projected to last for only forty-five years (down from 97 years in 1960). By contrast, coal reserves in the United States are projected to last for 300 years at current rates of production. The comparable life span of coal reserves for the Soviet Union is over 400 years and for the world as a whole is over 200 years. It is worth noting in this context that the world coal reserves, of which the Arab region has none, are close to five times the world reserves of oil in terms of energy content.

The oil link between the United States and the Arab oil producing countries is rather strong from the perspective of the United States as well as that of the Arab oil exporting countries. In 1970 the United States total imports of crude oil amounted to 1.32 million barrels per day (MBP) only. The Arab countries supplied 22 percent (291 thousand barrels per day) of United States' imports. It just so happened that these crude oil imports represented 21 percent of the exporting countries' total output of crude oil. Nine years later, in 1979, the United States found itself

---

*These countries are Algeria, Bahrain, Egypt, Iraq, Kuwait, Libya, Qatar, Saudi Arabia, Syria, and United Arab Emirates. Of these only the following are members of the Organization of Petroleum Exporting Countries (OPEC): Algeria, Iraq, Kuwait, Libya, Qatar, Saudi Arabia, and United Arab Emirates. There are six non-Arab members of OPEC: Ecuador, Gabon, Indonesia, Iran, Nigeria, and Venezuela. There are eleven Arab countries that are members of neither OAPEC nor OPEC: Djibouti, Jordan, Lebanon, Morocco, Mauritania, Oman, Somalia, Sudan, Tunisia, Yemen Arab Republic (North Yemen) and Peoples Democratic Republic of Yemen (South Yemen). Throughout this study only members of OAPEC are included in the term Arab region. See accompanying map.

Table 17.

UNITED STATES SHARE IN WORLD TOTAL OF ENERGY CONSUMPTION, 1965–1979

(in millions of barrels per day of oil equivalent, or MBDOE)

	Oil	Natural Gas	Coal	Hydro	Nuclear	Total
1965 Consumption						
United States	11.3	8.7	6.1	1.0	.1	26.9
World	31.1	13.0	31.0	4.7	.3	80.1
United States Share (%)	33.7	65.9	19.2	21.2	33.3	33.6
1973 Consumption						
United States	16.9	11.5	6.7	1.5	.4	36.6
World	56.5	21.6	33.6	6.6	1.0	119.3
United States Share (%)	29.6	53.0	19.9	22.7	40.0	30.7
1979 Consumption						
United States	17.9	10.0	7.7	1.6	1.5	38.1
World	64.1	26.1	39.7	8.3	3.1	141.3
United States Share (%)	27.9	38.3	19.4	19.3	48.4	27.0

importing 6.4 MBD of crude oil, with 47 percent of this amount or 3.0 MBD coming from the Arab world. This last amount, the 3 MBD, represented 15 percent of the Arab countries' total exports of crude oil. This pattern of trade relations between the United States and the Arab oil producing countries will be explored in more detail in the next part.

PART II

ARAB OIL AND
UNITED STATES' ENERGY NEEDS

Introduction

The relevance of oil to relations between the United States and Arab countries is by no means new. Although the importance of oil was highlighted by the events associated with and subsequent to the October 1973 war between Israel and Egypt-Syria, namely the Arab oil embargo and the considerable increases in the price of oil, the involvement of the United States in the petroleum history of the Arab World goes back to the turn of the twentieth century. This involvement has included every Arab oil producing country, from Algeria and Libya in North Africa to Saudi Arabia and Kuwait in the Arabian (Persian) Gulf region.

In the following pages a brief review of the history of United States involvement will be given and will be followed by an examination of the strong pattern of mutual economic dependence between the United States and the Arab World, a relationship that became the center of much controversy after 1973.

History

Soon after World War One the United States government urged the British government to allow American oil interests, mainly Exxon* and Mobil,† to have a share in the oil concession in Iraq. A concession is an agreement whereby the country grants to a company the exclusive right to explore and develop petroleum resources within a specific area. After several years of negotiations between the United States Department of State and the British Foreign Office, Exxon and Mobil were allowed to buy about one fourth in the Iraqi oil concession. This direct investment transaction was concluded in 1922 when these two corporations became the joint owner of 23.75 percent interest in the concession that had been obtained by Iraq Petroleum Company (IPC). IPC was formerly the Turkish Petroleum Company of pre-World War One days. The name was changed in 1929. The other owners of IPC represented British, Dutch, and French interests. The entry of Exxon and Mobil into the Arab World represented a major turning point in Arab-American energy relations.

After the restructuring of IPC another American company, Gulf Oil Corporation, was able to obtain, jointly with British Petroleum Company, a major concession in Kuwait. This was followed by another concession that was granted by Saudi Arabia to Standard Oil of California. This last concession was to be shared later on with Texaco, Exxon, and Mobil. Again, the owners of the IPC concession were able to obtain concessions in Qatar and the United Arab Emirates.‡ Here again Exxon and Mobil were holders of important shares of these concessions. These and other American oil companies (Occidental, Phillips, Getty, Continental, and others) had different shares,

*At the time Standard Oil Company (New Jersey).

†Originally Standard Oil Company of New York and reorganized as Socony-Vacuum before becoming Mobil Oil Corporation.

‡The United Arab Emirates (UAE) was formed in 1971 as a federation of seven princedoms (emirates). The most important of these seven in terms of oil resources and production is Abu Dhabi.

individually and collectively, in the oil concessions obtained in other Arab countries such as Libya, Algeria, Egypt, Syria, and Bahrain.

The significance of Arab oil for American oil interests can be appreciated if one keeps in mind that five of the seven largest oil companies in the world, sometimes referred to as the seven majors, are American. These five are Exxon, Mobil, Texaco, Gulf, and Chevron (Standard Oil of California). The other two major oil companies are British Petroleum and (Dutch) Shell. The importance of these seven oil companies can be appreciated by noting that in 1953 they were responsible for 87 percent of the oil produced outside the United States and the socialist countries. Given the fact that most of the oil that moves in foreign trade originates in the Arab region one can appreciate the importance of Arab oil for United States based multinational oil corporations, that is corporations formed by companies from different countries, and by extension for the American economy and the foreign policy objectives of the United States. And as late as 1972 the seven majors were producing 91 percent of the Middle East crude oil, that is oil before it is refined for specific uses. It is to be noted that the importance of Arab oil was not limited to the seven majors. It was also important to the well being of other multi-national oil corporations such as Occidental and Conoco.

It was mentioned earlier that United States oil companies had obtained oil concessions in all the oil producing countries in the Arab region. One of the most important features of these concessions was the fact that once they were obtained, governments had no control over the manner in which or the rate at which their oil wealth was exploited. In other words the management rights with respect to output and prices belonged to the companies alone. Another feature of the concessions was their duration in that company control was to last for decades. The third important feature was that in most cases especially in the Arab region these concessions covered almost the entire territory of the state. This fact hindered the state in its attempt to diversify the concession holdings in the hope of obtaining better terms. In exchange for all these privileges the state was to receive a prescribed payment per unit of output. This pattern of exploitation lasted until the early part of the 1970s when Iraq chose to nationalize the IPC concession. Nationalization increases the share of the host government in the concession. to anywhere between 51 and 100 percent. Other countries did likewise, with different forms of contractual arrangements with the oil companies operating in their countries.

Significance of Arab Oil To the United States*

The importance of Arab oil to the United States goes beyond its contribution to the energy needs of the United States, although this by itself is of paramount importance. In 1970 Arab oil accounted for only 2 percent of United States oil consumption (see Table 18 and Figure 9). This proportion jumped to almost 18 percent by 1977. But if we relate Arab oil to total oil imports by the United States we find that Arab oil provided almost 17 percent in 1960 and 36 percent by 1979. It is clear from these data that Arab oil provides a critical portion of United States oil needs. The importance of oil for the well being of any modern economy cannot be overstated because there is no substitute for oil in certain parts of the energy market. Oil played another critical role in the overall structure of American interests during the postwar period. This importance derived from the control which United States based oil companies had over

*See footnote, Part I, p.23, for a listing of Arab oil countries and for an explanation of the term Arab region used in this and subsequent sections.

ARAB OIL AS A PERCENT OF U.S. OIL CONSUMPTION AND U.S. OIL IMPORTS, 1960-1979

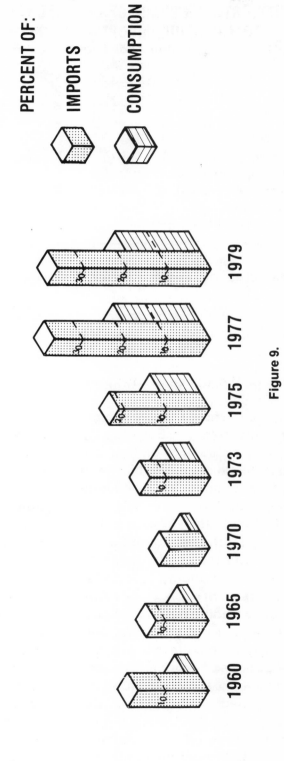

Figure 9.

Table 18.

UNITED STATES SHARE OF OIL IMPORTS FROM THE ARAB REGION TO TOTAL OIL IMPORT AND CONSUMPTION, 1960–1979
(in million barrels per day, or MBD)

	1960	1965	1970	1973	1975	1977	1979
Total Imports	1.8	2.5	3.4	6.3	6.1	8.8	8.3
From Arab Countries	.3	.3	.3	.9	1.4	3.2	3.0
Ratio of Arab Oil to Total Imports	16.7	12.0	8.8	14.3	22.2	36.7	36.1
United States Domestic Consumption	9.7	11.3	14.4	16.9	15.9	17.9	17.9
Ratio of Arab Oil to Total Consumption	3.1	2.7	2.1	5.3	8.8	17.9	16.7

Source: United States Department of Energy, *Annual Report to Congress, 1979.*

Arab oil by virtue of the concession agreements.

As earlier tables have shown the economic growth of Japan and Western Europe was greatly helped by the availability of Arab oil. The United States' interests in the viability and the growth of these economies is beyond dispute. It was one of the most important objectives of the United States' foreign policy to see prosperous and strong economies in the countries of Western Europe and in Japan. In order for the American economy to prosper and to maintain high rates of economic growth it was essential that its trading partners should have prospering economies of their own. This mutually reinforcing growth pattern resulted in one of the longest periods of economic growth in the market economies of the world under the leadership of the United States. It is in the attainment of these particular United States' policy objectives that Arab oil made its most important contribution. Without the continued availability of Arab oil to Western Europe and Japan the economies of these regions would

have been entirely different from what they are now. As a result the American economy would have been deprived of the vitality of these growing economies.

The economic benefits which Western Europe and Japan were able to derive from the availability of Arab oil were strengthened by the United States own policy toward imported oil. The oil import quota system which the United States put in place in 1957 limited oil imports to about one tenth of oil demand in this country. The main reason for this restrictive import policy was to insulate the American oil industry from the relatively low cost Arab oil. But by insulating its economy from oil imports the United States contributed to a downward pressure on the prices of Arab oil. This is so because oil resources in many countries were developed by American oil companies to meet the needs of the American market. This situation which lasted until the early part of the last decade was bound to change as soon as the American oil industry found itself unable to meet the constant rise in the demand for oil.

This fact led the United States to become first a major oil importing country and then the largest oil importing country in the world. By the time these developments made the United States dependent on foreign oil, it was inevitable that competition with other major oil importers would lead to upward adjustment in the price of oil. In other words the major price changes of the 1970s represent, among other things, a violent reaction to a number of policies and practices that had the effect of keeping the price of oil low and which could be traced to the decade of the 1950s.

Another aspect of the significance of Arab oil is related to the growth of major companies. In the process of contributing to the overall objective of having prosperous trading economies the United States also helped some of its own largest multinational corporations to grow. This observation is based on the fact that in the process of supplying the economies of Japan and Western Europe with a cheap source of energy these companies found themselves deriving an even larger ratio of their earnings from their foreign operations. And at the base of all this was Arab oil which the American oil companies were able to sell throughout the world. Oil company profits from foreign operations were brought back to the United States, thus helping the American balance of payments at a time when there was considerable movement of capital from the United States to be invested abroad. Investment in Arab oil fields proved to be highly profitable with rates of return on invested capital ranging up to 60 and 70 percent per year. This meant that oil companies were able to recover their invested capital in less than two years. Any expansion in their operations in the Arab region or elsewhere was financed from their earnings in the Arab region.

The profitability of oil investment in the Arab region may be illustrated by the purchase by Exxon and Mobil for 40 percent in the Aramco* concession which was held jointly by Chevron and Texaco. The original concession was obtained in the 1930s by Chevron, a concession that was shared later on with Texaco. These two owners agreed in 1946 to restructure the ownership of Aramco, the concession holder in Saudi Arabia, in order to allow Exxon and Mobil to buy equity shares in Aramco. Equity refers to the value that remains after subtracting mortgage and liability. The new arrangements gave Exxon a 30 percent interest in Aramco and gave Mobil a 10 percent interest. For the combined 40 percent share in Aramco the buying companies paid only $125 million. It goes without saying that this amount was recovered several times over by Exxon and Mobil.

Oil in the Context of United States' Energy Requirements

It was indicated earlier that oil was first produced in the United States in 1860. The United States was for a long time the largest oil producing, consuming, and exporting country in the world. Although it is now one of the three largest producing countries (with the Soviet Union and Saudi Arabia), it is still by far the largest oil consuming country in the world. Meanwhile its position has changed from being the largest exporter to the largest importing country.

It can be seen from the data on the United States' energy structure (Tables 19 and 20, and Figure 10) that the relative importance of oil in the total energy consumption increased between 1965 and 1979. In 1965 oil provided 41.5 percent of total United States' energy consumption. By 1979 the share of oil in the energy market had increased to 46.3 percent. Viewing the importance of oil from a different perspective one can see that while the United States' total consumption of energy increased by 42 percent between 1965 and 1979 the

*Aramco stands for the Arabian American Oil Company, a name that came into being on January 31, 1944.

Table 19.

UNITED STATES STRUCTURE OF ENERGY CONSUMPTION BY SOURCE, 1965–1979
(in millions of barrels per day of oil equivalent, or MBDOE)

	1965	1973	1979	Percent Average Annual Rate of Change 1965–73	1973–79
Oil	11.3	16.9	17.9	6.2	1.0
Natural Gas	8.7	11.5	10.0	4.0	– 2.2
Coal	6.1	6.7	7.7	1.2	2.4
Water Power	1.0	1.5	1.6	6.3	.8
Nuclear Power	.1	.4	1.5	37.5	45.0
Total	27.2	37.0	38.7	4.5	.1

Source: United States Department of Energy, *Annual Report to Congress, 1979.*

Table 20.

UNITED STATES ENERGY CONSUMPTION PERCENTAGE DISTRIBUTION BY MAJOR SOURCE, 1965–1979

	1965	1973	1979
Oil	41.5	45.7	46.3
Natural Gas	32.0	31.1	25.8
Coal	22.4	18.1	19.9
Water Power	3.6	4.1	4.1
Nuclear Power	.4	1.0	3.9
Total	100	100	100

Source: United States Department of Energy, *Annual Report to Congress, 1979.*

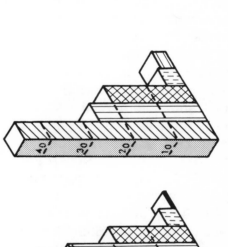

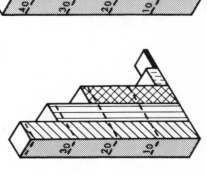

UNITED STATES OIL CONSUMPTION: PERCENTAGE DISTRIBUTION BY MAJOR SOURCE, 1965, 1973, and 1979

 OIL

 NATURAL GAS

 COAL

 HYDRO POWER

 NUCLEAR ENERGY

1979

1973

1965

Figure 10.

consumption of oil increased by 58 percent during the same period. This meant that the increase in the demand for energy was met by increasing the consumption of oil. This is reflected in the fact that oil was called upon to meet 57 percent of the increase in the demand for energy. Since production of oil from domestic sources could not keep up with the increase in the demand, it was inevitable that foreign oil had to be imported to fill the gap that was widening between rising demand and lagging domestic output. It should be noted that the discovery of Alaskan oil helped to increase domestic production but the output was not sufficient to lessen the importance of imported oil.

It is convenient to classify the energy market into four sectors, or categories. These are residential and commercial, industrial, transportation, and electric utilities. The relative importance of the sectors in the United States has changed over time. From the data in Tables 21 and 22 (see also Figure 11) it can be seen that the residential and commercial sector's direct use of energy increased by 38 percent between 1960 and 1979 while that of the industrial sector increased by 29 percent during the same period. Energy consumption by the transportation sector increased by 88 percent, and that of electric utilities by close to 200 percent. Total energy consumption during the same period went up by 77 percent. In terms of oil the sector with the largest consumption was transportation. This sector consistently consumed over one half of all the oil consumed in the United States. This high proportion is not surprising in light of the diversified and sophisticated transportation system in the country.

Two significant changes in the consumption of oil by sector are worth noting. The first relates to the growth in the industrial use of oil. This development is directly related to the increased use of oil as a feedstock for many industries. The other changes is related to consumption of oil by the electrical utilities sector. Note that this increase is basically of an intermediate rather than final-use nature; that is, the production of oil-based electricity is intended for consumption by the other two sectors (residential-commercial and industrial).

Value of Arab Oil in the Context of Total Imports

One of the most important aspects of Arab oil during the period before 1973 was the fact that it was inexpensive relative to prices of non-Arab oil of similar quality. The price of Arab oil was so low that it competed with American oil within the United States' market even after adding the cost of transportation from the Arab region to the American markets. For instance, in 1960 the landed cost (i.e., price of oil plus transportation cost) of foreign oil in the United States averaged $2.41 per barrel. The cost of domestic oil at the well-head in that year was $2.88. The margin of 47¢ per barrel increased to 57¢/barrel in 1970 when the cost of domestic oil increased to $3.18/barrel while that of imported oil increased only to $2.61/barrel. Even as late as 1973 the cost of imported oil was $3.58/barrel compared with the domestic cost of $3.89/barrel. Obviously, foreign oil had a comparative advantage over domestic oil; that is it had an edge over domestic oil because of its lower price. A comparison between the two sources of oil is given in Table 23. The oil industry in the United States became concerned enough that the American government found it necessary to restrict the inflow of foreign oil as early as 1957. The restrictive measures took the form of a quota on oil imports equal to 10 percent of domestic consumption. That is, the amount of imported oil could not exceed 10 percent of total consumption. These measures had the effect of slowing down the inflow of Arab oil. They were repealed in 1973 when they had lost their effectiveness as a result of the continued need of the American economy for foreign oil.

The data in Table 23 indicate several things. In the first place they show that the value of imported Arab oil was insignificant throughout the decade of the 1960s and up to 1973. In both absolute terms and relative to the total imports

Table 21.

UNITED STATES ENERGY CONSUMPTION PERCENTAGE DISTRIBUTION BY SECTOR, 1960–1979

	1960	1970	1979	Percentage Change 1960–1979
Residential and Commercial	24.7	17.2	19.2	37.6
Industrial	33.3	29.2	24.4	29.3
Transportation	23.8	23.9	25.3	87.3
Electric Utilities	18.6	24.4	31.2	198.0
Total	100	100	100	77.3

Source: United States Department of Energy, *Annual Report to Congress, 1979.*

Table 22.

UNITED STATES OIL CONSUMPTION: VOLUME AND PERCENTAGE DISTRIBUTION BY SECTOR, 1960–1979

	1960		1970		1979		Percentage Change 1960–1979
	MBD	%	MBD	%	MBD	%	
Residential and Commercial	2.6	26.5	3.5	23.8	3.5	19.0	34.6
Industrial	1.9	19.4	2.5	17.0	3.7	20.1	94.5
Transportation	5.1	52.0	7.8	53.1	9.7	52.7	90.2
Electric Utilities	.3	3.1	.9	6.1	1.6	8.7	433.3
Total	9.8	100	14.7	100	18.4	100	87.8

Source: United States Department of Energy, *Annual Report to Congress, 1979.*

UNITED STATES OIL AND ENERGY CONSUMPTION: PERCENTAGE DISTRIBUTION BY SECTOR, 1960, 1970, and 1979.

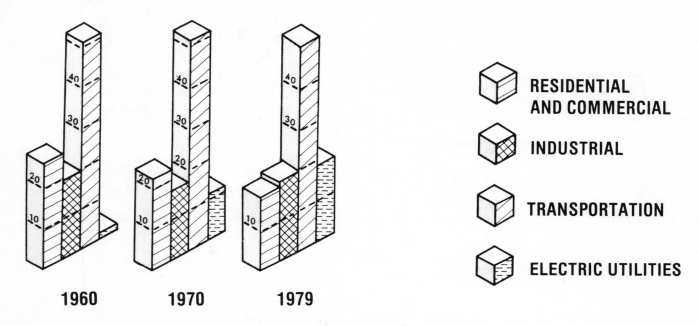

Figure 11.

Table 23.

UNITED STATES VOLUME, VALUE, AND UNIT PRICE
OF IMPORTED OIL, 1960-1979

	1960	1970	1973	1979
Total Petroleum Imports (MBD[*])	1.8	3.4	6.3	8.4
Imports from Arab Region (MBD[*])	.3	.3	.9	3.0
Share of Arab Oil to Total Imports (percent)	16.7	8.8	14.3	36.1
Value of Total Petroleum Imports ($ billion)	1.6	2.7	7.7	56.5
Value of Arab Oil ($ billion)	.3	.2	1.1	20.4
Value of Total United States Imports ($ billion)	22.5	61.0	89.4	244.2
Share of Arab Oil Imports to Total Imports (percent)	1.3	.3	1.2	8.4
Landed Cost per Barrel ($)	2.41	2.61	3.58	18.62
United States Domestic Oil Price ($ per barrel)	2.88	3.18	3.89	14.27

[*]MBD = million barrels per day

Source: United States Department of Energy, *Annual Report to Congress, 1979; International Economic Report of the President, 1975.*

of the United States Arab oil imports remained within 1 percent of total imports. In absolute terms the value of imported Arab oil increased from $0.3 billion in 1970 to $1.1 billion in 1973. This tripling of the cost was due primarily to a tripling in the quantity of imported oil which had increased from .3 million barrels per day (MBD) in 1970 to .9 MBD in 1973. The significant increase in the landed cost of imported oil in 1979 reflects two major changes. First the volume of oil imports increased from 6.3 MBD in 1973 to 8.4 MBD in 1979, an increase of 2.1 MBD or 33 percent. But oil imports from the Arab region increased from .9 MBD in 1973 to 3.0 MBD—an increase of 2.1 MBD or of 233 percent. In other words the entire increment in this country's oil imports came from the Arab region. The other factor responsible for the increase in the cost of imported oil was the steep rise in the price of oil that started in

October, 1973, within the context of the October War between Israel and Egypt-Syria. Although crude oil prices were either frozen or increased gradually between 1974 and 1978 they increased dramatically in 1979 and 1980. This is shown in the changes that characterized the rise in the landed cost of imported oil. In 1975 the average landed cost of imported oil was $12.45 per barrel. In 1976 it was $13.34 per barrel and by 1977 it had reached $14.31, to rise again to $14.38 a barrel in 1978. By 1979 the cost had reached an average of $18.62 per barrel.

Foreign Oil and
United States' Energy Policy

Historically the United States has relied on sources of oil within the Western Hemisphere to meet its requirements of imported oil. These sources include Canada, Venezuela, and Mexico. It was natural for the United States to draw its imported oil from these sources because of their proximity to the centers of consumption and because of the presence of American oil interests in both Canada and Venezuela. But as more American oil firms found it to be profitable to have oil investment in the Arab region it was natural that these firms would import Arab oil to meet their own needs from these new sources of supplies. The decisions of these firms to import Arab oil into the United States was aided by the fact that the cost of Arab oil was so low that it was possible for it to compete in the United States' energy markets with the much higher cost of oil from the Western Hemisphere. Moreover, the non-major multinational oil companies (the independents) were very much interested in having their own sources of raw material (crude oil) without having to be at the mercy of the majors. These facts led to two developments. The first was the increase in the number of independents seeking oil arrangements in the Arab region as well as in other parts of the world. Secondly, the inflow of Arab and foreign oil gave rise to concern that imported oil might constitute a threat to the United States' oil industry. This concern prompted the introduction of the quota system mentioned earlier. Aside from the restrictive quota system the United States did not have a particular energy policy with respect to foreign oil. The overall objective of policy was to insure that the United States trading partners would continue to meet their needs for oil at low prices. A related objective was that these oil supplies be provided to the maximum extent possible by United States based oil firms.

The October 1973 war has been referred to a few times in the discussion. Its relevance to the question of oil supply and oil prices is as follows. In that month Egypt and Syria simultaneously attacked Israeli forces that had been occupying Egyptian and Syrian territory since June 1967. Egypt and Syria had not been able to remove the Israeli occupation by other means. The United States immediately rushed to the aid of Israel with a massive airlift of weapons and supplies. This enormous assistance to a country that was occupying Arab lands made Arabs especially angry with the United States, and in fact they regarded the United States as a party to the Israeli occupation and to the persistence of the occupation. But Arab countries had hardly any options that could be exercised. The importance of Arab oil to American needs loomed large, and Arab oil exporting countries decided to halt the selling of their oil to the United States. This action came to be known as the Arab oil embargo, and it lasted till early in 1974. Whereas the halt in the flow of oil was short lived, a much more enduring effect was a galloping increase in oil prices that continued all through the 1970s. These events prompted the United States to look into ways and means for adopting an energy policy. The overriding objective was to develop mechanisms that would cushion the economy from an unexpected oil shock, be it in the form of price change or of an unforeseen interruption of supply.

The policy dilemma in a situation like this is the conflict between two objectives. The first objective was to assure the flow of needed oil to meet an ever expanding demand. The other was to attempt to lessen dependence on foreign supplies. What compounded the problem for policy makers was the inability of the Western Hemisphere sources to make up for any serious shortage of supply should such a shortage ever develop again. Moreover, the higher prices of imported oil did not stop the United States economy from continuing its upward demand for foreign oil. Whereas imported Arab oil decreased from .9 million barrels per day in 1973 to .8 MBD in 1974, Arab oil imports increased to 1.4 MBD in 1975, to 2.4 MBD in 1976, and to 3.2 MBD in 1977. This meant that any reduction in oil imports would have adversely affected an economy that was already suffering from stagnation and inflation.

Although many observers of the energy scene did not think that a serious threat to oil imports was very likely, the Nixon administration advocated what came to be known as Project Independence. The thrust of the program was to recommend certain policy measures that would make the country energy self sufficient. Analysts of the energy situation doubted that self-sufficiency could be achieved, or that it was even necessary to adopt such a policy. Political and economic realities proved that the critics were correct. One of the major problems with President Nixon's Project Independence was its projected heavy reliance on the expansion of nuclear energy. Since the contribution of this particular source of energy had always been overestimated there should have been no reason to expect that its projected contribution after 1973 would be attainable. It is important to note that both environmental and health (radioactivity) hazards associated with this particular source of energy have led the American and other societies not to accelerate their reliance on nuclear energy. The failure of Project Independence led to less ambitious energy policies that were intended to slow down the rate of reliance on imported fuels. These policies had two major objectives. First was the promotion of conservation through better insulation, retrofitting, and improved automobile mileage performance. The second was to encourage the production of oil from domestic sources through tax and price incentives, and also to promote the production of other sources of energy such as coal and natural gas.

One of the first important pieces of legislation aimed at reducing the dependence of the United States on foreign oil was the 1974 Energy Supply and Environmental Coordination Act. This act authorized the federal government to require major fuel burning installations to convert from oil to coal, the latter being in abundant domestic supply. In 1975 the Energy Policy Conservation Act was enacted to accomplish, among other things, two objectives. One of these objectives was to stimulate domestic output of crude oil. The other was to allow, over a period of forty months, domestic prices to rise gradually to the

level of world market prices. By allowing gradual price increases the Act attempted to protect the American economy from the then higher world oil prices. By the same token the gradual increase in domestic prices provided an incentive for marginal and high cost oil to be produced. Furthermore this act introduced several conservation measures such as minimum automobile mile per gallon requirements, a weatherization program, and a plan for gasoline rationing if needed. In 1976 the Energy Conservation and Production Act was passed with the intent to expand and strengthen the commitment of the government to conservation.

In addition to these policy measures that were designed to promote conservation and to increase the production of domestic sources of energy, the United States was at the same time playing a leading role at the international level. This role manifested itself in at least three major ways. The first was the creation of the International Energy Agency (IEA). The United States together with other industrialized countries agreed to establish the IEA to coordinate certain energy measures among member countries with the specific aim at reducing their dependence on imported oil. The most important of these measures relate to the exchange of information, the buildup of oil stockpiles, and the sharing of oil supplies should there be a reduction in supplies below a certain level. The oil sharing system was considered the most important provision of the agreement in that it was intended to minimize the impact of any major interruption on the economy of any one country. The provisions involved and the system devised to put them into effect are very difficult to evaluate since the system has not been put to test. The second other major foreign energy policy pursued by the United States was to participate in the deliberations of the Conference on International Economic Cooperation (CIEC), known as the North-South dialogue. The term North-South refers to developed and less developed countries. The United States attempted soon after the oil price revolution of 1973–74 to coordinate with other industrialized countries negotiating with the oil producing countries about oil prices and supplies. The Organization of Petroleum Exporting Countries (OPEC)* insisted that other developing countries should participate in the deliberations and that these deliberations and negotiations should relate not only to energy but also to other issues and concerns that were important to these developing countries.

The economic problems that plagued most Third World countries stemmed in many cases primarily from the long history of having been subjected to European colonialism, imperialism, and exploitation for over two hundred years. These systems of exploitation helped the industrialized countries to engage in a long and sustained process of wealth transfer to themselves from the countries of Asia, Africa and Latin America. The countries of the Third World attempted, under the auspices of the United Nations, to solve some of their problems through the introduction of structural changes in the functioning of the international economy. The attempts to have these changes introduced and the various resolutions of the United National General Assembly (from 1974 to 1976) to have the changes implemented came to be known as the New International Economic Order (NIEO). The position of OPEC and its Arab member countries (whose exports of crude oil amount to over 70 percent of OPEC oil exports) was that the issues of NIEO should be discussed and that representatives from Third World countries should play an active part. Although the position of the United States was one of opposition to the inclusion of Third World countries in the North-South dialogue it was compelled, however, to change that position since the United States was the only industrial country to hold to that position. Negotiations lasted for two years, but the results of the CIEC were marginal and of only

educative significance as regards any of the major issues involved.

The failure of CIEC coincided with a new energy policy thrust that was embarked upon by the Carter Administration which came to office in 1977. This new posture is the third major initiative taken by the United States at the international level. President Carter introduced a doctrine which implied that any threat to the free flow of oil from the Middle East would be construed as a threat to the United States not only in terms of its national interests but also with respect to its national security. In his January 23, 1980 State of the Union Address President Carter said: "Let our position be absolutely clear. An attempt by any outside force to gain control of the Persian Gulf region will be regarded as an assault on the vital interests of the United States of America, and such an assault will be repelled by any means necessary, including military force."† It was thus to be understood that the United States was prepared to resort to military action to keep the oil flowing from the Arab region. This proclaimed policy was considered to be the motivating force behind the decision to create the rapid deployment force for the purposes of military intervention in the Arab region.

At the domestic level Carter's energy policy was contained in the National Energy Plan (NEP) announced in April 1977. NEP marked a significant change from the energy policies of the previous administrations in that it considered conservation measures as the most important source for developing "additional energy." Based on a price and tax structure, the NEP sought both to offset dependence on foreign oil through a massive increase in domestic coal production and to bring United States' crude oil and natural gas prices to world levels in order to encourage conservation and exploration. However, environmental concerns limited the projected

*See footnote, Part I, p. 23.

†United States Department of State Bureau of Public Affairs, *Current Policy*, No. 132.

contribution of nuclear energy and coal in reducing United States' dependence on foreign oil. In addition, Congress failed to adopt Carter's 1977 NEP. Instead, it enacted the Natural Gas Policy Act of 1978 which embodied important price increases for this source of energy. In an attempt to deal with possible crude oil supply disruptions, the United States government also began stockpiling crude oil in 1978 through the Strategic Petroleum Reserve (SPR) program which was initially incorporated in the 1975 Energy Policy Conservation Act. While the SPR provided for storage of up to one billion barrels of crude oil, by 1980 this figure had not yet been reached.

The Iranian Revolution of 1979 had the effect of raising oil prices throughout the world. The disparity between world prices and domestic prices led President Carter to announce in April 1979 that controls over oil prices would be removed over a period of time extending from June 1979 to September 1981. Included in this policy was a provision for the federal government to tax a major portion of the profits that would accrue to oil companies as a result of the decontrol program. And out of this tax the federal government was to fund in part the synthetic fuel program. Although the final phasing out of the price control was to be completed in September 1981, President Reagan accelerated the decontrolling process when he announced in February 1981, the elimination of all price controls.

Arab-American Trade and Finance Relations

A comprehensive understanding of energy relations between the United States and the Arab region requires an understanding of another set of relations. The interaction between the United States and the Arab region has other important dimensions such as trade in merchandise other than oil, and trade in services such as banking, consulting, technology transfer, shipping, tourism, etc. It also takes the form of investing Arab funds in the money and capital markets of the United States. This financial interaction assumed an increasing importance after 1973.

In terms of merchandise trade, oil alone is responsible, as to be expected, for most of the Arab region's exports to the United States. In 1979 oil imports from the Arab region represented over 90 percent of that region's exports to the United States (see Table 24 and Figure 12). From the data in Table 24 it can be seen that the merchandise trade grew considerably. In 1970 exports from the Arab region to the United States amounted to only $200 million. By 1979 the value of exports had increased to $22.5 billion. Similarly the Arab region's imports from the United States increased considerably, from $530 million to $11.1 billion during the same period (Table 25 and Figure 13). The share of the Arab region with the United States increased considerably if compared with the rest of the world. In 1970 exports from the Arab region to the United States accounted for less than 2 percent of that region's total exports. By 1979 the United States had received over 16 percent of the Arab region's exports. Imports from the United States in the Arab region increased from less than 10 percent of the region's total imports in 1970 to close to 14 percent in 1979. In terms of overall United States' merchandise imports the contribution from the Arab region increased from one half of one percent in 1970 to 10.3 percent in 1979 (Table 24).

Table 26 shows the development of trade between the Arab region and the United States during most of the 1970s (see also Figure 14). In 1973 and 1974 the United States had a small merchandise surplus with the Arab region. By 1975 the surplus had changed to a deficit of $1.7 billion. This deficit continued to increase until it reached $11.3 billion in 1979. The cumulative deficit for the period 1975 to 1979 amounted to $31.2 billion. As was indicated earlier trade covers not only goods such as oil but also services. The export and import of services between the Arab region and the United States consistently reflected a surplus in favor of the United States. It can be seen from

Table 24.

ARAB REGION: TOTAL EXPORTS AND EXPORTS TO THE UNITED STATES, 1970–1979
($ billion)

	1970 U. S.	1970 World	1972 U. S.	1972 World	1976 U. S.	1976 World	1979 U. S.	1979 World
Algeria	.01	1.01	.11	1.29	2.35	5.33	4.97	9.39
Iraq	N	1.02	.02	1.21	.11	8.48	.61	19.00
Kuwait	.03	1.65	.05	2.56	.07	9.84	.09	16.48
Libya	.04	2.36	.18	2.31	2.27	8.31	5.04	15.06
Qatar	---	.22	N	.40	.41	2.21	.27	3.62
Saudi Arabia	.02	2.42	.22	4.52	1.75	36.13	8.97	56.33
United Arab Emirates	.07	.55	.02	1.08	1.13	8.51	1.96	13.49
Bahrain	.01	.24	.08	.20	.12	1.52	.01	2.04
Egypt	.02	.76	.01	.83	.06	1.52	.38	2.35
Syria	N	.20	N	.29	.01	1.07	.17	1.59
Total	.20	10.43	.69	14.69	8.28	82.92	22.47	139.35

	1970	1972	1976	1979
Share of Exports to United States to total Exports (percent)	1.9	4.7	10.0	16.1
United States Total Imports ($ billion)	42.43	58.88	129.57	218.93
United States Imports from Arab Region ($ billion)	.20	.69	8.28	22.47
Share of United States Imports from Arab Region to Total Imports (percent)	.5	1.1	6.4	10.3

N: Negligible

Source: International Monetary Fund, *Direction of Trade Annual, 1980*; *International Financial Statistics Yearbook, 1980.*

UNITED STATES — ARAB TRADE

AS PERCENT OF TOTAL TRADE, 1970 - 1979

U.S. IMPORTS FROM SELECTED
ARAB COUNTRIES
AS PERCENT OF TOTAL U.S. IMPORTS

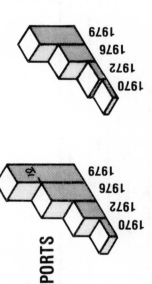

Figure 12.

EXPORTS OF SELECTED
ARAB COUNTRIES TO U.S.
AS PERCENT OF TOTAL EXPORTS

UNITED STATES — ARAB TRADE BY COUNTRY, AS PERCENT OF TOTAL TRADE OF THE ARAB COUNTRY, 1970 - 1979.

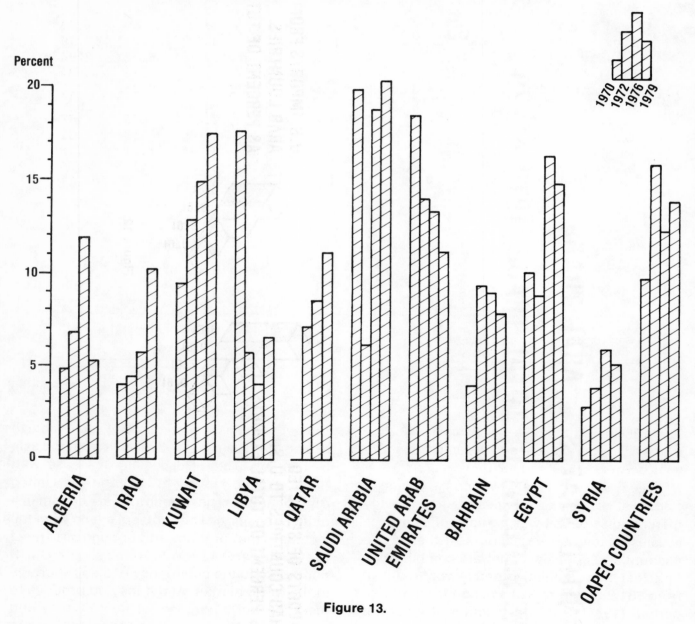

Figure 13.

Table 25.

ARAB REGION: TOTAL IMPORTS AND IMPORTS FROM THE
UNITED STATES, 1970-1979
($ billion)

	1970 U. S.	1970 World	1972 U. S.	1972 World	1976 U. S.	1976 World	1979 U. S.	1979 World
Algeria	.06	1.26	.10	1.49	.63	5.34	.44	8.42
Iraq	.02	.51	.03	.72	.22	3.90	.49	4.79
Kuwait	.06	.63	.10	.78	.49	3.32	.84	5.70
Libya	.10	.56	.06	1.05	.13	3.21	.52	7.96
Qatar	---	.07	.01	.14	.07	.82	.15	1.34
Saudi Arabia	.14	.71	.70	1.14	1.63	8.69	5.36	26.27
United Arab Emirates	.05	.27	.07	.50	.46	3.42	.76	6.78
Bahrain	.01	.25	.03	.32	.15	1.66	.18	2.31
Egypt	.08	.79	.08	.90	.62	3.81	1.21	8.18
Syria	.01	.36	.02	.54	.14	2.36	.17	3.36
Total	.53	5.41	1.20	7.57	4.54	36.53	11.12	80.11

Share of United States in Total Imports (percent): 9.8   15.9   12.4   13.9

Source: See source for Table 24.

Table 26 that United States export of services to the Arab region exceeded its imports every year between 1973 and 1979.

If we combine these two components of trade, i.e. goods and services, we find that the merchandise deficit was partially offset by the surplus in the service account. If the service account and goods account are combined we find that the cumulative deficit was reduced from $31.2 billion to $22.4 billion for the same period (1973-1979). In addition to current account transactions (goods and services), there are the financial transactions that have to be dealt with in order to have as complete a picture as possible. The nature of some of the Arab economies is such that they are unable to use all the revenues they receive from the sale of their oil. This is especially the case with Saudi Arabia, Libya, Kuwait, and the United Arab Emirates. These countries are caught in a difficult position. The problem stems from the fact that if they produce only enough oil to meet their current revenue needs, they will contribute to the emergence of an oil shortage in the world market. And if they do produce to meet the world demand for their oil they find themselves with funds that cannot be used in their economies at the present time.

The disparity between the absorbtive capacity of these Arab oil producing countries

Table 26.

ARAB REGION: GOODS AND SERVICES EXPORTS TO AND IMPORTS FROM THE UNITED STATES, 1973–1979
($ billion)

	1973		1974		1975		1976		1977		1978		1979	
	EX	IM	EX	IM	EX	IM	EX	IM	EX	IM	EX	IM	EX	IM
Algeria	.21	.24	1.10	.51	1.29	.90	2.35	.63	3.01	.62	3.36	.41	4.97	.44
Iraq	.01	.05	---	.19	.02	.37	.11	.22	.38	.22	.24	.35	.61	.49
Kuwait	.05	.15	.05	.22	.88	.43	.07	.49	.18	.66	.09	.61	.09	.84
Libya	.31	.10	---	.11	1.37	.14	2.27	.13	3.88	.20	3.63	.29	5.04	.52
Qatar	.02	.02	.08	.03	.06	.05	.41	.07	.30	.12	.20	.12	.27	.15
Saudi Arabia	.38	.38	1.10	.49	1.14	.72	1.75	1.63	3.94	2.73	5.98	4.26	8.97	5.36
United Arab Emirates	.07	.13	.37	.23	.68	.42	.46	.15	1.54	.58	1.04	.65	1.96	.76
Bahrain	.02	.04	.13	.08	.25	.09	.12	.15	.14	.13	.06	.14	.01	.18
Egypt	.02	.02	.01	.01	---	.76	.06	.62	.03	.79	.10	1.10	.38	1.21
Syria	---	.02	---	.04	---	.10	.01	.14	.04	.12	.10	.10	.17	.17
Subtotal	1.09	1.25	2.84	2.29	5.69	3.98	8.28	4.54	13.44	6.17	14.80	8.03	22.47	11.12
Services	.13	.07	.23	1.08	.66	1.78	1.04	2.34	1.91	3.09	2.47	3.63	2.90	6.05
Total	1.12	1.32	3.07	3.37	6.35	5.76	9.32	6.88	15.36	9.26	17.27	11.66	25.37	17.17
Balance	-.20		-.30		+.59		+2.44		+6.10		+5.61		+8.20	

Source: See source for Table 24.

UNITED STATES — ARAB TRADE IN GOODS AND SERVICES AS PERCENT OF TRADE TOTAL, 1973 - 1979

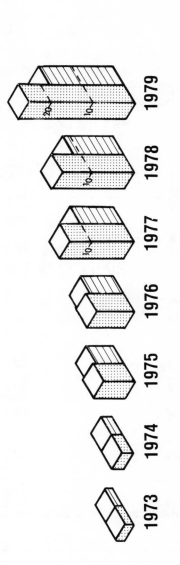

ARAB REGION'S

EXPORTS

IMPORTS

1973 1974 1975 1976 1977 1978 1979

Figure 14.

and the actual revenue they receive from the sale of oil has given rise to what is known as current account surplus. And after having disbursed part of this surplus to international organizations or to borrowing countries, what is left is called petrodollars or OPEC surplus. This surplus is simply the funds that are deposited by certain oil producing countries in the money and capital markets of the United States and Europe.

Although it is difficult to trace each country's investment activity, studies show that a major part of the surplus funds tend to gravitate to the money market in the United States. It can be seen from Table 27 (see also Figure 15) that between 1974 and 1979 the Arab oil producing countries with surplus funds invested an estimated total of $72 billion in the United States. Comparing this amount with the accumulated deficit (on goods and services) of $22.4 billion we find that the flow of funds to the United States is more than three times the deficits that the country incurred with the Arab region.*

SUMMARY 1973–1979

1. United States cumulative merchandise deficit with the Arab region was $31.2 billion (Table 26).

2. United States cumulative goods and services deficit with the Arab regions was $22.4 billion (Table 26).

3. Arab cumulative investment in the United States was $72 billion (Table 27).

4. Estimated net transactions in favor of United States amounted to $49.6 billion (72 minus 22.4).

Table 27.

ARAB REGION: ESTIMATES OF INVESTMENT OF CURRENT ACCOUNT SURPLUS IN THE UNITED STATES, 1974–1979
($ billion)

	Total Investment	Investment in United States
1974	41	9
1975	23	6
1976	30	10
1977	25	7
1978	11	2
1979	44	38
Total	174	72

Source: Derived from OAPEC, *Secretary General Seventh Annual Report, 1980* (Arabic), Kuwait, 1981.

*It is very interesting to note that the United States balance of trade deficit (merchandise export and import) with Japan alone was $46 billion for the period 1973 to 1979. This compared with a total deficit with the Arab region of $31.2 billion for the same period.

INVESTMENT BY ARAB REGION OF CURRENT ACCOUNTS SURPLUS

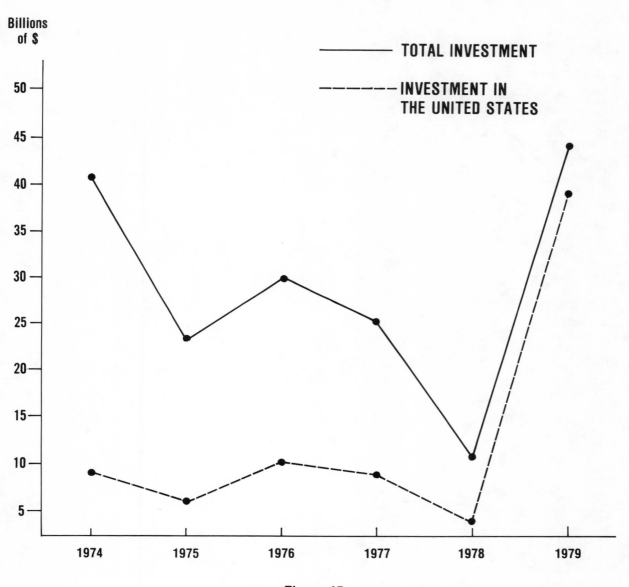

Figure 15.

PART III

OIL AND ECONOMIC CHANGE
IN THE ARAB REGION

Introduction

In addressing the question of oil and economic change in the Arab region* it is important to state that only seven countries are responsible for 96 percent of the oil produced by Arab countries. The remaining Arab countries have either small quantities of oil to export such as Egypt, Oman, Syria, and Tunisia, or none at all. Oil importing Arab countries include Jordan, Lebanon, Morocco, Sudan, Yemen, Mauritania, and Somalia. In this section economic change in the Arab region is considered, and the focus will be on the seven major Arab oil exporting countries (Algeria, Iraq, Kuwait, Libya, Qatar, Saudi Arabia, and the United Arab Emirates). Factors that will be examined are: the importance of oil as a source of foreign exchange in these countries, oil's importance as a source of revenue for the state, the relative importance of the oil sector in the gross national product, the life span of oil reserves, and some of the effects which oil has brought to the Arab region at large.

The Changing Importance Of Oil in National Economies

Historically the oil sector in each and every Arab oil producing country was developed by foreign enterprises. As was seen in Part I the instrument for the development of oil resources in these countries was the concession system. The system was simple in that the state gave to the companies its rights for the development of its oil and the setting of the price of oil in exchange for a fixed sum of money per unit of

output. This payment was about 25 cents per barrel prior to 1950, and it was not related to the market price of oil. The major change affecting the revenue from oil occurred in the fifties when a change in the terms of the concessions allowed governments to levy tax on company profits. Thus a government's revenue per barrel of oil increased with a company's profit, whereas in the earlier system the government did not benefit from price increases. This change resulted in a revenue increase from 75¢ to 80¢ per barrel between 1950 and 1970. Another major change was the accelerated development of oil resources in the Arab region, first in Southwest Asia (Iraq, Kuwait, Saudi Arabia, Qatar, United Arab Emirates) and later in North Africa (Libya and Algeria). Libya became an oil exporting country in 1961, the United Arab Emirates in 1962, and Algeria in 1968. Both Egypt and Syria started their oil exports in the 1970s.

The combination of these two changes set the stage for a sustained growth in Arab oil output and exports for the following two decades. At the same time there took place the increase in the importance of the oil sector in the economies of the oil producing countries.

Oil and the Gross National Product

The increase in oil output and revenue created an important change in the relation of this sector to the other sectors of the economy. The importance of the oil sector increased over

*For definition see footnote, Part I, p. 23.

the years with the increase in output and the per barrel revenue. Oil's importance began to assume even a greater significance after the structural changes that shifted control over pricing and output decisions from the oil companies to the oil producing countries. In 1970, the last year during which oil companies had the sole control over prices, the combined oil revenue for five countries (Algeria, Iraq, Kuwait, Libya, and Saudi Arabia) was $4.8 billion. In that year the total output of these countries amounted to 12.6 million barrels per day, almost all of which was exported. The revenue per barrel in that year averaged 83¢.

The relative importance of the oil sector to the economy can be measured by relating oil revenue to the gross national product (GNP). In 1970 the contribution of oil revenues to the combined GNP of the five countries mentioned above was 21 percent. It can be seen from the data in Table 28 that the relative importance of the oil sector varied considerably, ranging from 7 percent in Algeria to 39 percent in Kuwait. In 1974, as output continued to increase and prices were raised dramatically, the revenue

from oil in seven countries (the five already mentioned plus Qatar and the United Arab Emirates) reached a level of $53.4 billion. The aggregate (or combined) GNP for this group of seven countries reached $78.4 billion in 1974. These changes meant that the importance of the revenue from oil had increased to 68 percent of the combined GNP of these countries, a considerable increase over 1970. The data in Table 29 (see also Figure 16) show that the oil sector contributed 28 percent of the Algerian GNP and 97 percent of the GNP of Saudi Arabia in 1974. The role of the oil sector in other countries varied between two these rates. By 1979 the contribution of the oil sector had increased to 73 percent of the combined GNP of these countries. In that year the revenue from oil was $135 billion, while the aggregate GNP was $184 billion.

Oil and Foreign Trade

Like most Third World countries Arab oil resources were developed to meet the needs

Table 28.

SELECTED MAJOR ARAB OIL PRODUCING COUNTRIES: OIL OUTPUT EXPORTS, OIL REVENUE AND GROSS NATIONAL PRODUCT, 1970

	Output MBD	Export MBD	GNP $ billion	Oil Revenue $ billion	% of GNP
Algeria	1.0	1.0	4.6	.3	7
Iraq	1.5	1.5	2.9	.5	18
Kuwait	3.0	2.6	3.0	.9	39
Libya	3.3	3.3	5.7	1.3	23
Saudi Arabia	3.8	4.2	3.9	1.2	31
Total	12.6	12.6	20.1	4.2	21

Source: World Bank, *Trends in Developing Countries* (1973); *Middle East Economic Survey*, various issues.

Table 29.

MAJOR ARAB OIL PRODUCING COUNTRIES: GROSS NATIONAL PRODUCT, OIL REVENUE, OIL OUTPUT, AND OIL EXPORTS, 1974 and 1979

	GNP $ billion		Oil Revenue $ billion		Ratio of Oil Revenue to GNP (percent)		Oil Output		Oil Export		Population Million		Per Capita GNP $	
	1974	1979	1974	1979	1974	1979	1974	1979	1974	1979	1974	1979	1974	1979
Algeria	11.8	28.9	3.3	7.5	28	26	1.0	1.2	.9	1.1	16.3	19.1	770	1580
Iraq	10.6	30.4	5.7	21.3	54	70	2.0	3.5	1.9	3.3	10.8	12.8	984	2410
Kuwait	10.9	21.9	8.6	16.9	79	77	2.5	2.5	2.2	2.4	1.0	1.3	11727	17270
Libya	11.9	23.4	6.0	15.2	50	65	1.5	2.1	1.5	1.4	2.3	2.9	4900	8210
Qatar	2.0	3.8	1.8	3.6	82	95	.5	.5	.5	.5	.2	.2	10270	16590
Saudi Arabia	23.3	62.6	22.6	57.5	97	92	8.5	9.5	7.9	9.2	7.0	8.1	2842	7370
United Arab Emirates	7.9	13.0	5.5	12.9	70	99	1.7	1.8	1.4	1.7	.5	.8	16122	15590
Subtotal	78.4	184.0	53.4	134.9	68	73	17.8	21.3	16.3	19.6	38.1	45.5	2058	4043
Bahrain	.6	2.1	.3	.8	33	38	.1	.1	---	.1	.2	.3	2310	5460
Egypt	10.4	18.6	.1	1.2	10	6	.1	.5	---	.3	36.4	41.0	286	460
Syria	5.6	7.8	.4	1.1	7	14	.1	.2	.1	.1	7.1	8.1	788	1070
Total	95.0	212.5	54.3	138.0	57	65	18.1	22.1	16.4	20.1	81.8	94.9	1161	2239

Source: OAPEC, *Secretary General Annual Report* (various issues); World Bank, *World Bank Atlas, 1977 and 1980.*

56

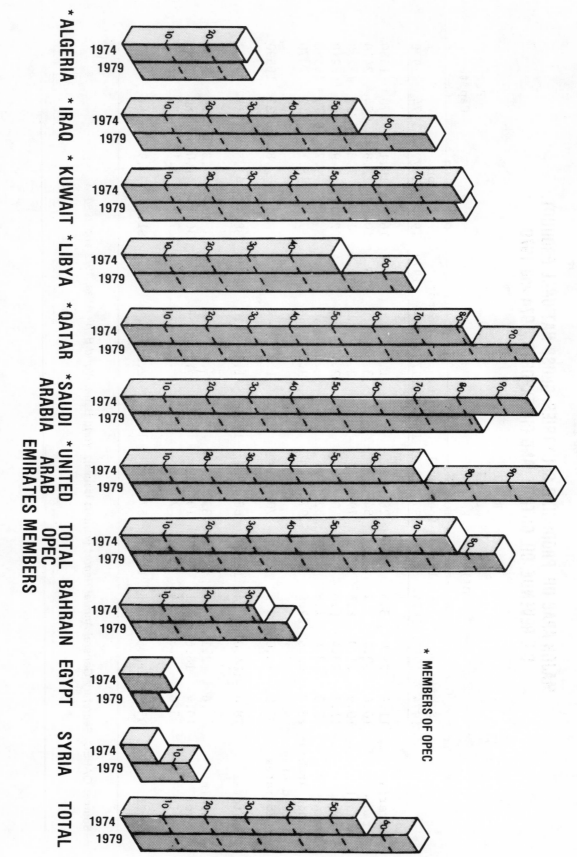

OIL REVENUE AS PERCENT OF GNP BY ARAB COUNTRY, 1974 and 1979

Figure 16.

not of the domestic economies but those of foreign countries. In the case of oil, as was noted earlier, most of it flowed to the economies of the industrial countries, and it has continued to do so. Since most of the countries had very few resources other than oil, it was only natural for oil to be either the dominant or practically the only export of these countries. The rise in the importance of oil exports for countries that have other exports can be seen from the data in Table 30 and Figure 17. Libya and Saudi Arabia depend almost exclusively on oil as their sole export. The exports of countries like Kuwait, Qatar, and the United Arab Emirates also are dominated almost entirely by oil. Syria and Egypt were much less dependent on oil for their exports in earlier years. By 1979 these two countries became more dependent on oil exports, which accounted for 48 percent of Egypt's exports and 71 percent of Syria's. As to Algeria and Iraq it can be seen from the data that as time went on they, too, became increasingly dependent on oil exports. In 1970 export earnings from oil accounted for 70 percent of Algeria's total export earnings. By 1979 the figure was 91 percent. For Iraq the ratio had increased from 90 percent in 1970 to 99 percent in 1979. For the seven largest oil exporting countries the rising importance of oil in the 1970s transformed them into one raw material exporting countries.

This dependence on oil export earnings has the consequence of exposing these countries to the effects of economic and policy changes that may take place outside their boundaries. Thus, any change in the rate of economic growth in oil importing countries is bound to affect the volume of oil exports, a change that may or may not be in harmony with the national interests of the oil exporting countries. Oil importing countries may introduce policy changes that influence their oil imports, and this in turn influences the course of economic development in oil exporting countries. Another aspect of this dependence on oil exports is that it tends to create another dependency, namely high levels of imports that

can be supported by only higher oil export earnings. The tremendous increases in these countries' imports of goods and services can be seen from the data in Table 31 and Figure 18.

In 1974 these seven countries imported a total of $26 billion. By 1979 their imports had jumped to $97 billion. In other words, imports had increased at an average annual rate of 54 percent during the five year period. No other country or group of countries had experienced such a high rate of import growth during this period. The implications of this jump will be dealt with later. Suffice it to say for now that the increase in export earning made it possible to increase imports in a way that could not have been supported by earnings from sources other than oil.

Arab Oil: Some Economic And Social Ramifications

Like all other developing countries, Arab oil exporting countries suffer from a state of underdevelopment with all its manifestations of short life expectancy, high rate of illiteracy, distorted economic structures, maldistribution of income, low rates of economic growth, inadequate labor force, and heavy dependence on the markets of industrial countries. While a critical factor in the process of economic development is the availability of capital (that is, wealth that can be used for the production of more wealth) capital is not the only factor. It is not enough for purposes of economic development simply to have capital. Other factors include the size of the economy itself, the diversification of the economy, the availability of an adequate and appropriate labor force, and conditions and institutions that are helpful to development.

In spite of these common features with other developing countries Arab oil exporters found themselves in a unique position due to the special role of oil in their economies. This is so because the sharp increase in the revenue from oil in the 1970's created opportunities, but also severe economic and social problems and stresses.

Table 30.

MAJOR ARAB OIL EXPORTING COUNTRIES:
TOTAL EXPORTS AND OIL EXPORTS, 1970-1979

($ billion and percent)

	1970			1974			1979		
	Total Exports $	Oil Exports $	Ratio of Oil Percent	Total Exports $	Oil Exports $	Ratio of Oil Percent	Total Exports $	Oil Exports $	Ratio of Oil Percent
Algeria	1.0	.7	70	4.7	4.4	94	8.2	7.5	91
Iraq	1.0	.9	90	6.6	6.5	98	21.5	21.3	99
Kuwait	1.7	1.6	94	10.1	10.0	99	18.1	16.6	91
Libya	2.4	2.4	100	7.2	7.2	100	15.2	15.2	100
Qatar	.2	.2	100	2.0	2.0	100	3.8	3.6	95
Saudi Arabia	.2	.2	100	30.7	30.7	100	57.6	57.6	100
United Arab Emirates	.6	.6	100	6.5	6.5	98	13.6	12.8	94

Source: Derived from OAPEC, *Secretary General Seventh Annual Report* (Kuwait, 1981).

OIL EXPORTS AS PERCENT OF TOTAL EXPORTS BY ARAB COUNTRY, 1970, 1974, and 1979.

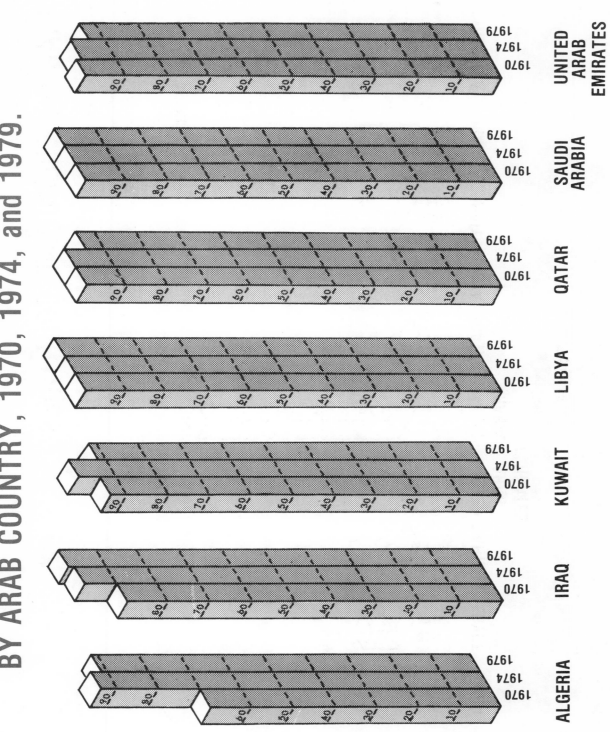

Figure 17.

ARAB OIL PRODUCING COUNTRIES: IMPORTS OF GOODS AND SERVICES, 1974 and 1979

*MEMBERS OF OPEC

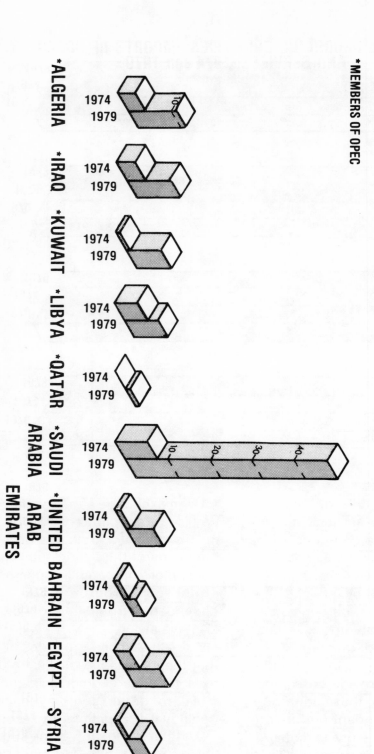

Figure 18.

Table 31.

ARAB OIL PRODUCING COUNTRIES: IMPORTS OF GOODS AND SERVICES, 1974 and 1979

($ billion)

	1974	1979
Algeria	4.7	12.1
Iraq	4.7	10.6
Kuwait	1.0	7.7
Libya	5.4	8.3
Qatar	.3	1.3
Saudi Arabia	8.1	48.7
United Arab Emirates	1.7	7.0
Subtotal	25.9	95.7
Bahrain	1.1	2.5
Egypt	3.7	8.2
Syria	1.4	3.8
Total	32.1	110.2

Source: OAPEC, *Secretary General Seventh Annual Report, 1980* (Kuwait, 1981).

The sudden increase in the oil wealth resulted in the first place in a sharp rise in inflationary pressures in these countries. This was to be expected due to the inability of the domestic economies to increase the supply of goods and services to meet the oil induced rise in demand. This in turn led to a sharp increase in the volume and the costs of imports, a fact which added to the strength of inflation.

Given these facts and the fact that governments were called upon to provide or expand social programs and services, these countries found themselves more dependent on foreign goods and services to meet the rise in demand both in the public and the private sectors. This increased dependence on imports led in turn to higher degree of dependency on oil since oil was the only commodity whose export could be increased in order to generate the necessary

foreign exchange earnings that were needed to pay for foreign imports.

This increased reliance on imports to meet the needs of the public and private sectors for consumer and capital goods had the negative effect of perpetuating a pattern of oil dependency from which these countries tried to move away.

As to the social problems, it suffices to say that the increase in oil wealth provided incentives to large segments of the population to move to urban centers hence undermining traditional occupations and economic activities.

Another dimension of the sudden increase in the oil wealth was the emergency of certain groups that were able to derive considerable incomes from foreign trade, land speculation and all types of public spending. The rise of these groups tended to widen the inequality in the pattern of income distribution a fact that tended to give rise to social, political and economic alienation on the part of those who could not derive similar benefits from oil.

It is important to note that for some Arab oil exporters a relatively new dilemma was presented by the sudden increase in oil wealth. For those countries whose internal absorbtive capacity tended to lag behind their earnings the dilemma was how to manage balance of payments surpluses in a world economy that has been characterized by inflation and currency fluctuations. In other words the capital surplus countries found the value of their foreign assets being eroded by inflation and currency depreciation at a time when they were called upon to deplete their national wealth at a faster rate than the needs of their national economies would have required. It should be mentioned in this connection that much has been said about the fact that the per capita income in some of the Arab oil producing countries are among the highest in the world. And they are. In 1979 the figures for Kuwait, Qatar and the United Arab Emirates were in fact the highest in the world—higher than for Switzerland, Germany or the United States. But this comparison is misleading, for at least two reasons. The higher per capita income in some of the Arab oil producing countries does not reflect an income derived by factors of production as is the case in most other countries. It is not an income although it is classified as such. It is actually the price paid for the exchange of one form of asset for another. It is, in other words, the proceeds that accumulate as a result of the sale of an asset. It is similar to the sale of a piece of land. The owner of the land can rent the land and derive an income as long as he has a title to the land, or he can sell the land. In the first case the land was used as a factor of production that commanded an income. In the second case the land represented wealth that was converted into cash. The proceeds from the sale of the land can either be spent on consumption or can be converted into another income producing factor of production that will yield income year

after year. The situation of the oil producing countries is similar to that of the landlord. By producing the oil and selling it they have the option of either spending the proceeds from the sale of their natural wealth on consumption, public and private, or they can convert their oil into income-producing physical factors of production, such as factories, harbors, roads, farms, schools, and hospitals. They may prefer to leave the money in the bank and use the earnings for current consumption. A country may choose a combination of all these options. An alternative to all or to some of the options would be for the country to reduce its output of oil on the premise that oil in the ground is more valuable in the future than oil produced at the present time.

It should be remembered that not all Arab oil producing countries are at the same stage of economic development, that not all are producing more wealth than can be integrated in their economies, and that they are not all endowed with high levels of oil reserves that would enable them to meet the cost of their imports of goods and services. An example is Egypt, considered to be one of the poorest countries in the world with a per capita income of $390 in 1978. In that year there were only thirty-eight countries with lower per capita incomes. Syria had a per capita income of $930 in 1978. This figure was much less than the average of the 54 countries that the World Bank classified as the middle income countries in that year; that average was $1250. Both Algeria and Iraq belong also to this group of countries. Algeria's per capita income in 1978 was $1260 while that of Iraq was $1860. Bahrain, also in this group, had a per capita income of $1500 in that year. Note that the countries that have the highest per capita incomes of all the Arab oil producing countries are those with the smallest populations. They are Kuwait, Qatar, and the United Arab Emirates. Two other countries, Libya and Saudi Arabia, produce and export more oil than is needed to meet their domestic and international obligations. Therefore they continue to have a surplus in the balance of payments, meaning that they make more money on what they sell than they spend on what they buy. These five countries are what is known as the capital surplus countries. As was noted in the first part these countries are responsible for the bulk of the so-called OPEC* surplus fund.

In all major oil exporting countries the sudden increase in the income from oil created certain severe economic and social problems. One of the earlier problems which every one of these countries had to face was the sudden increase in inflation rates. It was only natural that the increase in revenue from oil should be used to finance both capital and consumer goods on a large scale. This was made necessary by the needs of economic development and by the desire to meet consumer needs that had not been met for so long. Such needs included basic food necessities that were taken for granted in the truly affluent societies. The accelerated pace of development created and/or raised personal income which in turn increased the demand for goods and services that either were not available or were available in a limited supply. This fact helped push domestic prices up. The domestic inflation was compounded by the fact that the infrastructure (physical and permanent installations that facilitate production, transportation, communication, and the overall management of goods and services) in each and every one of these economies was incapable of handling the influx of the imports, a fact that accentuated the shortage and contributed to inflation.

The requirements needed for the development prompted the governments of countries that have a labor deficit but a capital surplus to allow and encourage the influx of non-citizens from other countries in the region and from beyond the region. The higher salaries in the capital surplus and underpopulated countries attracted a large number of workers of all categories—skilled, semi-skilled, and professional. The main centers of attraction were Kuwait, Saudi Arabia, Qatar, United Arab Emirates, and to a smaller extent Iraq. The workers came primarily from Jordan, Palestine, Yemen, Lebanon, and Egypt. Other labor exporting countries included India, Pakistan, Bangladesh, Sudan, and Korea. As with every major movement of labor there are both economic benefits and economic and social costs. The most obvious benefit to the labor importing country is that its needs for particular skills are met immediately and at no capital investment.

While a labor importing country derives certain benefits from the inflow of foreign labor the same thing can be said about the labor exporting countries. One of the obvious benefits is the reduction in unemployment in the labor exporting country in the early stages of the outflow of labor.

Another benefit to the labor exporting country is the increase in the inflow of funds from abroad as the migrant workers send part of their earnings to their countries of origin. In its *1981 World Development Report* the World Bank has estimated that in certain cases foreign exchange earnings from this particular source i.e. remittance inflows, have exceeded the value of the merchandise exports of some labor exporting countries. Thus in 1978 Jordan's remittance inflow was estimated to be 175% of that country's exports. In the case of both Yemen Arab Republic and People's Democratic Republic of Yemen remittance inflows constituted almost the only source of foreign exchange earnings. Nor were these flows confined to Arab countries only. Thus in the case of Pakistan remittance inflows were equal to 93% of that country's exports. The comparable ratios for Bangladesh and India were 21% and 18% respectively.

It should be noted that the increase in the cost of imported energy created certain balance of payments problems to many developing countries. It is important to note, however, that the higher cost of imported

*See footnote, Part I, p. 23.

energy was not the only cause for these difficulties. The continued worldwide inflation and economic slowdown in industrial countries compounded problems of developing countries in more than one way. The nature of the international economic system is such that most of the trade of developing countries is with industrial countries. This means that every increase in prices in industrial countries is transmitted to developing countries through the imports by the latter from the former. And whenever there is a recession or a sluggish rate of growth in the industrial countries the developing countries tend to suffer as a result of a reduction in their exports to the industrial countries. In addition to the fact that trade with industrial countries is resonsible for the major share of problems of payments and debt in developing countries, there is another issue that tends to confuse the role of imported oil in the problems of developing countries. Of 124 oil importing developing countries, eight (Brazil, India, Hong Kong, Korea, Phillipines, Thailand, Turkey, and Singapore) account for over 70 percent of all the oil imported. In 1978 these eight countries imported 3 of the 4 million barrels per day imported by all the developing countries. With the exception of India, Phillipines, and Thailand, the other five countries are far richer and more industrially advanced than most of the Arab oil exporting countries.

OPEC countries, especially the Arab members, have responded to these new difficulties burdening the developing countries by channeling aid funds to them. The aid so provided has been directed either through the OPEC Fund for International Development or through a member country's own national fund for external development. Although aid had been directed to developing countries before 1973, it was not until later in the 1970s that aid programs were institutionalized. According to data prepared by the World Bank, the official development assistance from OPEC member countries to other developing countries ranged from $5.6 billion in 1976 to $4.3 billion in 1978. The bulk of the aid was provided by the Arab

member countries. It can be seen from data in Table 32 and Figure 19 that in comparison with aid provided by industrial countries, Arab aid is several times higher in terms of the respective gross national products of the two groups.

It should be noted in this connection that data on Arab aid tends to underestimate the real magnitude of the aid. The reason for this is that conventional national account methodologies do not take into consideration the fact that the revenue from oil is not an ordinary income. Revenue from oil represents the sale value of a finite capital asset. In other words had GNP accounts given consideration to the depletable nature of oil, the ratio of aid to GNP would have been higher than the data indicate.

The special nature of oil revenue was recognized by Mr. McNamara, former president of the World Bank, when he suggested in 1974 that GNP data for oil producing countries should be adjusted by a depletion factor of 30%. Applying this depletion factor to the data on Arab aid (Table 32) would result in an upward revision in the GNP percentages from 5% to 6.5% in 1975; 4% to 5.2% in 1976 and 1977; from 2.6% to 3.4% in 1978 and from 2.4% to 3.1% in 1979.

Oil and the Future of The Arab Economies

Although the 1970s was to Arab oil producing countries a major landmark in the history of control over their oil resources, it should be remembered that the oil sector had always played a major role in the economic development of the Arab region. Its contribution was, to be sure, an indirect one in that it provided a source of revenue which the governments used to finance development projects. Development budgets and plans were prepared in every Arab oil producing country with the aim of using the revenue from oil to restructure the economy in order to lessen dependence on the oil sector. The degree of success in this endeavor varied, depending on the particular economic, political, and social

Table 32.

OFFICIAL DEVELOPMENT ASSISTANCE FROM MAJOR ARAB OIL PRODUCING COUNTRIES AND INDUSTRIAL COUNTRIES, 1975–1979

($ billion and as percentage of Donor GNP)

	1975		1976		1977		1978		1979	
	$	% of GNP	$	% of GNP	$	% of GNP	$	% of GNP	$	% of GNP
Iraq	.2	1.7	.2	1.4	.1	.2	.2	.8	.9	2.9
Libya	.3	2.3	.1	.6	.1	.7	.2	.9	.1	.6
Saudi Arabia	2.0	5.4	2.4	5.7	2.4	4.3	1.5	2.8	2.0	3.2
Kuwait	1.0	8.1	.6	4.4	1.5	10.6	1.3	6.4	1.1	5.1
Qatar	.3	15.6	.2	8.0	.2	7.9	.1	3.4	.3	5.6
United Arab Emirates	1.0	14.1	1.1	11.0	1.2	10.2	.7	5.6	.2	1.6
Total Above Countries	4.9	5.0	4.7	4.0	5.5	4.0	3.9	2.6	4.6	2.4
United Kingdom	.9	.4	.9	.4	1.1	.5	1.5	.5	2.1	.5
Sweden	.6	.8	.6	.8	.8	1.0	.8	.9	1.0	.9
Germany	1.7	.4	1.6	.4	1.7	.3	2.3	.4	3.4	.4
Japan	1.1	.2	1.1	.2	1.4	.2	2.2	.2	2.6	.3
France	2.1	.6	2.1	.6	2.3	.6	2.7	.6	3.4	.6
United States	4.2	.3	4.4	.3	4.7	.3	5.7	.3	4.6	.2
All Industrial Countries	13.8	.4	13.8	.3	15.7	.3	20.0	.3	22.3	.3

Source: World Bank, *World Development Report, 1980.*

66

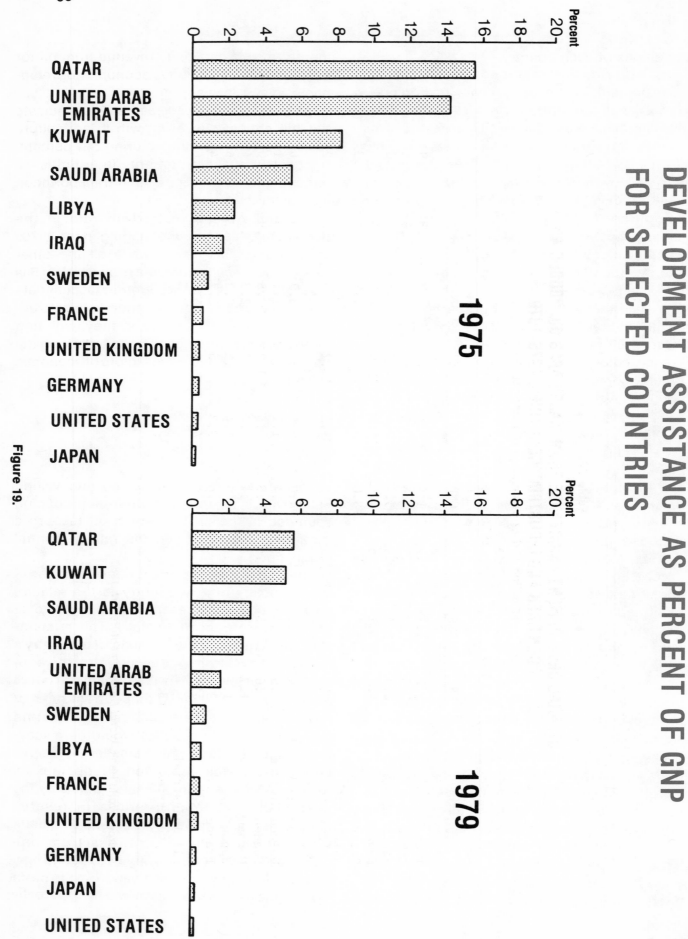

Figure 19.

DEVELOPMENT ASSISTANCE AS PERCENT OF GNP
FOR SELECTED COUNTRIES

conditions of each country.

In Algeria, for instance, the focus of development in the early years of independence was on education and vocational training. This was followed by investment in various industrial projects and heavy investment in the energy sector. The average annual rate of growth in the manufacturing sector in the period 1970 to 1977 was almost 7 percent compared to a rate of 17.7 percent in construction (the comparable rate in the 1960s was 1.3 percent) and 13.5 percent in electricity and gas (compared to a negative growth rate, that is a decrease, of 1.5 percent in the 1960s). The main reason for the low rate of growth in construction and utilities in the 1960s was that these sectors were under the control of the French colonial rule for most of that decade. Algeria continued to expand its development budget during the 1970s by raising its share in total government spending from 39 percent in 1973 to 45 percent in 1980.

In Iraq heavy emphasis was placed on manufacturing, construction, public utilities, and transportation. As can be seen from the data in Table 33, in the 1970s all these sectors achieved growth rates that were much higher than the growth rates of the 1960s. These development accomplishments were not accompanied by similar development in the agricultural sector. On the contrary agriculture exhibited negative growth rates or actual declines, in the 1970s. Development plans in both Algeria and Iraq, as well as in Libya, have been aimed at pouring a considerable amount of investment capital into all sectors of the economy, in the hope of increasing the production rates of the various sectors.

Data on Kuwait's attempt in economic development in the 1960s indicate that public policy was targeted to accomplish two important objectives. The first objective was to build the country's infrastructure. And the other was to provide social services primarily in the fields of health, education and housing. Kuwait not only succeeded in meeting these goals but it also pioneered a new experiment in international aid giving in that it was the first

country to use part of its revenue from oil for the purposes of regional economic development.

The data for the decade of the 1970's indicate that the annual rate of growth in agriculture was 6.1 percent, in manufacturing 24.7 percent, in construction 13 percent, in utilities 12 percent, and 12 percent also in transportation and communications.

In Saudi Arabia the growth rate in the manufacturing sector was lagging in the 1970s compared to the 1960s, while all the other sectors showed considerable growth rates. It is expected that as these countries accumulate experience and as their economic infrastructures become more mature they will find themselves in a better position to transform the oil wealth into broad based factors of economic growth.

Oil Reserves And Their Depletion

One of the most complicated and vexing issues that face Arab and other oil producing countries is the rate at which oil resources should be depleted, i.e., the rate of annual production.

One of the very important considerations in the oil policy of any country is the relation between crude oil reserves and the rate at which oil is produced annually. This relationship will determine the life span of a country's reserves. A country like Kuwait with reserves of 65 billion barrels of oil in 1979 would run out of oil in 76 years, assuming that current rates of production would be maintained and that no new significant oil reserves would be discovered. Algeria, with a much smaller oil reserve endowment, was projected to deplete its resources in less than 20 years, that is by the year 2000. These differences in reserve endowment and rate of output tend to dictate different policy considerations from one country to another. A country like Algeria may be expected to stress conservation and to profit from higher prices since its oil wealth apparently

Table 33.

ECONOMIC INDICATORS: SELECTED ARAB OIL PRODUCING COUNTRIES,
1960-1970 and 1970-1977
Average Annual Growth Rate (percent)

Sector	Algeria 1960-70	Algeria 1970-77	Iraq 1960-70	Iraq 1970-77	Saudi Arabia 1960-70	Saudi Arabia 1970-77	Kuwait 1960-70	Kuwait 1970-77
Agriculture	.4	.2	5.7	- 1.5	1.0	3.7	n.a.	6.1
Manufacturing	7.7	6.9	5.9	11.5	10.7	4.8	n.a.	24.7
Construction	1.3	17.7	5.7	23.1	8.2	23.5	n.a.	13.1
Electricity, Gas, and Water	- 1.5	13.5	12.3	15.6	17.2	12.9	n.a.	12.2
Transportation and Communications	0.1	7.9	4.4	14.3	14.5	19.0	n.a.	12.1
Administration and Defense	7.2	5.7	8.8	10	9.0	9.6	n.a.	n.a.
Mining	19.3	2.5	4.0	10.6	11.1	13.1	n.a.	-11.5
Trade and Finance	.8	2.1	9.6	10.8	11.2	15.5	n.a.	7.3
Other Sectors	-25.8	17.4	9.5	9.5	7.1	6.9	n.a.	7.7
Gross Domestic Product	4.6	5.3	6.1	10.8	9.9	12.7	-	0.1

Source: World Bank, *World Tables*, 1980.

will be depleted in a short period of time. Another country like Saudi Arabia can afford to produce beyond its current financial needs since its reserves have a much longer life than those of Algeria.

Looking at all the major Arab oil producing countries one must be impressed by the speed with which these countries have been depleting their oil wealth. It can be seen from Table 34 and Figure 20 that in 1960 oil reserves were projected to last 96 years at that year's rate of production. By 1970 the life span of the reserves was reduced to 61 years and by 1979 the life span was reduced again to 44 years.

One of the most revealing aspects of the data in Table 34 is the phenomenal rise in the size of the reserves. In 1960 these countries had a combined reserve total of 153 billion barrels. By 1979 the figure had more than doubled to 331 billion. Yet during the same period the life span of the reserves had been cut by more than 50 percent, from 96 to 44 years, indicating a very fast rate of depletion of this national wealth. The issues of conserving national wealth and meeting at the same time global needs for oil are intertwined. The role of Arab oil in the future of world energy needs will be examined in the next part.

Table 34.

MAJOR ARAB OIL PRODUCING COUNTRIES: RESERVES/OUTPUT RELATION, 1960–1979

	1960			1970			1979		
	Reserves Billion Barrels	Output MBD	Reserves Life Years	Reserves Billion Barrels	Output MBD	Reserves Life Years	Reserves Billion Barrels	Output MBD	Reserves Life Years
Algeria	5.2	.2	78	8.1	1.0	22	8.4	1.2	20
Iraq	27.4	1.0	72	32.0	1.5	57	31	3.5	24
Kuwait	65.6	1.7	105	80.0	3.0	73	68.5	2.5	76
Libya	---	---	---	29.2	3.3	24	23.5	2.1	31
Qatar	2.5	.2	39	4.3	.4	33	3.8	.5	20
Saudi Arabia	53.0	1.3	110	141.4	3.8	102	166.5	9.5	48
United Arab Emirates	---	---	---	12.8	.8	45	29.4	1.8	44
Total	154	4.4	96	307	13.8	61	331	21	44

Source: Ian Seymour, *OPEC: Instrument of Change* (London, Macmillan, 1980).

ESTIMATED YEARS OF OIL RESERVES FOR ARAB COUNTRIES, 1960, 1970, 1979

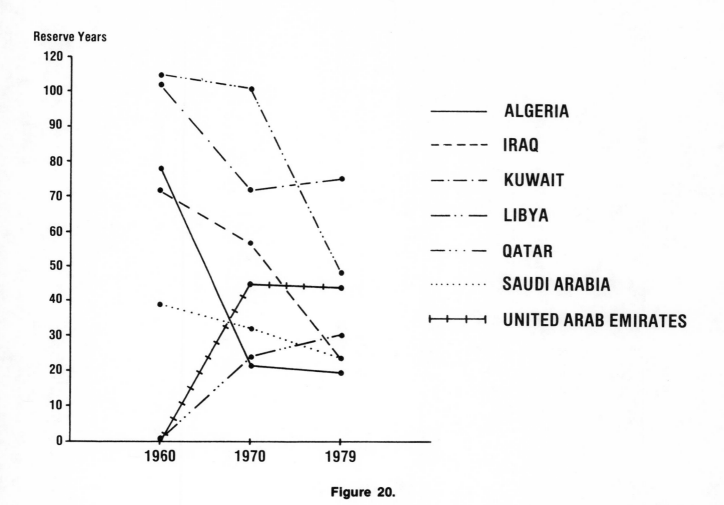

Figure 20.

PART IV

ARAB OIL IN THE
FUTURE OF WORLD ENERGY

Introduction

Any attempt to assess the importance of Arab oil within the context of future world energy calls for some knowledge of a number of factors. These include projections of energy supply and demand, projections of the changing mix of energy consumption by source, the changing need of the various economies for various sources of energy, and, of course, the policies of the Arab countries with respect to the conservation of their own natural resources. In this part an assessment of these and other forces will be undertaken in an attempt to place the Arab oil in the context of world energy future.

It should be noted at the outset that the Organization of Petroleum Exporting Countries (OPEC) will be referred to frequently since most projections and studies in this field tend to concentrate on member countries of OPEC rather than on Arab oil producing countries. This fact will not affect the analysis since the Arab countries that are members of OPEC produce the major share of oil produced by all members of OPEC, 72 percent in 1980. The Arab members of OPEC (Saudi Arabia, Iraq, Libya, Algeria, United Arab Emirates, Kuwait, and Qatar) also produce the bulk, 96 percent, of oil produced in the Arab region.*

Future Demand
For World Energy

The factors that are most important for making projections about demand for energy, whether the reference is to a country or to a region, are the following: population growth, rates of economic growth, the ratio of energy to the gross national product, prices asked for energy, and the ability of an economy to substitute one source of energy for another. During the decade of the 1970s, many studies and projections were published on the subject of future energy demand. Each of these studies had to be based on certain assumptions. The assumptions differed from study to study, and from year to year. Of course this meant that the projections also differed. Here is an example of how a change in the underlying assumption results in a change in the projected demand for energy. If we suppose that world economic growth will be 4 percent per year and that the energy/GNP ratio is 1.0, then it will take 18 years for the world demand for energy to double. But if we assume that the rate of economic growth will be 3.5 percent and that the energy/GNP ratio will be reduced 0.7, then it will take 29 years for the world demand for energy to double. The importance of the energy/GNP ratio can be viewed differently. For example, in the United States 6.5 barrels of oil equivalent were needed to produce $1000 of GNP in the period 1960 to 1969. The same $1000 of GNP is projected to need only 4.25 barrels of oil equivalent in the year 2000. This is a reduction of one third for the same amount of output. Similarly the higher cost of energy projected for the balance of this century will provide an incentive to substitute one source of energy for another and to stimulate develop-

*The other six members of OPEC are Nigeria, Indonesia, Iran, Venezuela, Gabon, and Ecuador. See footnote, Part I, pg. 23.

ment of domestic sources of supply. These two facts will tend to slow the rate of growth in the demand for imported oil to the extent that oil can be replaced by another or cheaper source of energy. As was indicated earlier there is a close link between the rate of economic growth and the demand for energy. This was clearly shown by the relatively high rates of economic growth and the high rate of demand for energy in all parts of the world in the decades of the 1950s and 1960s. Economic growth rates in the 1960s ranged from an annual average of 4.9 percent for the developed market economies (United States, Canada, Japan, Western Europe, Australia, and New Zealand) to 6.6 percent for the Centrally Planned Economies, and 5.3 percent for the developing countries of the Third World.

These growth rates are not expected to be matched in the period 1980 to 2000. The industrialized countries are expected to experience a growth rate of 2.5 percent per year, or about one half of the growth rate of the 1960s. The Centrally Planned Economies are also projected to have lower growth rates, 4.5 percent compared with 6.6 percent. The developing countries on the other hand will show considerable variations in their growth rates depending on whether they are high growth or low growth countries. For the first group the rate of economic growth is projected to average between 5 percent and 6 percent per year, while it would be about 3 percent for the second group. For the world as a whole, the real economic growth rate is projected to average 3 percent per year for the period 1980 to 2000. Given these lower economic growth rates and the gains in fuel efficiency mentioned earlier, it is not surprising to expect that demand for energy between 1980 and 2000 will be growing at much lower rates than in earlier years.

World total energy demand is projected to rise from 140 MBDOE (million barrels per day of oil equivalent) in 1980 to 225 MBDOE in 2000, or at an average growth rate of 2.5 percent per year compared with an average growth rate of 5.3 percent per year for the period 1965 to 1973. It can be seen from the data in Table 35 that for the industrial countries the growth rate in energy demand is projected to be 1.2 percent per year compared to an annual average rate of 5.2 percent for the period 1965 to 1973. The most drastic change within this group will be experienced by Japan

Table 35.

ENERGY DEMAND GROWTH RATES: PERCENT PER YEAR
BY MAJOR REGION, 1965–2000

	1965–1973	1973–1979	1979–2000
United States	4.3	0.8	0.8
Western Europe	5.1	1.5	1.5
Japan	11.4	1.4	2.1
Industrial Countries	5.2	1.2	1.2
Centrally Planned Economies	5.1	5.3	4.9
Developing Countries	6.9	5.3	4.9
World Total	5.3	2.9	2.4

Source: Exxon, *World Energy Outlook* (1980).

whose rate of growth of demand for energy is projected to decline from an average rate of 11.4 percent for the period 1965–73 to 2.1 percent for the balance of the century.

The high rate of growth in the demand for energy by Third World countries, 4.9 percent compared with 1.2 percent for the industrial countries, reflects three important conditions. First the population growth in the developing countries is expected to be much higher than that of the industrial countries. Second, all developing countries are expected to increase their demand for energy as the process of development and industrialization continues to proceed in the 1980s and 1990s.

This will be especially true of the few developing countries that will continue their efforts to expand their industries. These countries (for example Brazil, Argentina, Hong Kong, and South Korea) are responsible for the major portion of the energy consumed in all Third World countries. The third factor is that the OPEC countries themselves are expected

to be high energy users because of their efforts to industrialize and because of their wish to have energy based industries. These differences in the growth rates will change the shares of the major consuming regions in total energy consumption. It can be seen from the data in Table 36 that the demand for total energy by industrial countries will increase from 80 MBDOE in 1980 to 96 MBDOE in 1990 and to 109 MBDOE in 2000. During the two decades under consideration, energy consumption of this group of countries is projected to increase by 36 percent compared with an increase of 60 percent for the world as a whole. These countries' share in the world demand for energy is expected to decline from 57 percent in 1980 to 49 percent in 2000.

As to the Centrally Planned Economies their demand for energy is projected to rise from 42 MBDOE in 1980 to 55 MBDOE in 1990 and to 68 MBDOE in 2000. Their share in the world's total demand for energy is projected to remain about 30 percent of the world total throughout this

Table 36.

WORLD DEMAND FOR ENERGY BY MAJOR REGION, 1980–2000
(in millions of barrels per day of oil equivalent, or MBDOE and percent)

	1980		1990		2000	
	MBDOE	% of World Total	MBDOE	% of World Total	MBDOE	% of World Total
United States	38	27	41	23	45	20
Western Europe	27	19	33	18	38	17
Japan	8	6	10	5	12	5
Industrial Countries	80	57	96	52	110	49
Centrally Planned Economies	42	30	55	30	68	30
Developing Countries	18	13	33	18	48	21
World Total	140	100	185	100	225	100

Source: See source for Table 35.

period. The developing countries, including OPEC countries, are expected to increase their consumption from 18 MBDOE in 1980 to 48 MBDOE in 2000. This increase of 167 percent, compared with a 60 percent increase for the world as a whole, will raise their share of world energy consumption from 13 percent in 1980 to 21 percent in 2000.

Considering next the major sources of energy, it is projected that the demand for oil will increase from 66 MBD in 1980 to 70 MBD in 2000 (see Table 37 and Figure 21). The share of oil in total energy consumption will decline from 47 percent in 1980 to 35 percent in 2000. It is important to note that synthetic crude oil provides only 4 MBD in 1990 and only 9 MBD in 2000. The contribution of natural gas to total energy consumption is projected to increase from 27 MBDOE in 1980 to 43 MBDOE in 2000. Its share of total energy demand, however, will remain about 19 percent. Coal's share in energy consumption will increase from 26 percent in 1980 to 28 percent in 2000. Actual consumption is projected to increase from 36 MBDOE in 1980 to 63 MBDOE. Another way of looking at coal is to compare it to oil. In 1980 coal consumption was 54 percent of total oil consumption. In 2000 coal consumption is projected to rise to 80 percent of oil consumption.

The source of energy that will register the highest rate of growth is nuclear energy. From only 3 MBDOE in 1980 consumption will increase to 22 MBDOE in 2000. Nuclear energy's share of total energy consumption will rise from 2 percent to 10 percent during the same period.

Finally, the contribution of hydro power will increase from 6 percent to 8 percent or from 8 MBDOE to 18 MBDOE between 1980 and 2000.

Future Supply
Of World Energy

The question of energy supply is similar to the question of demand in that all projections are subject to change depending on the assumptions used. These assumptions include cost of production, exploration of areas, both

Table 37.

WORLD ENERGY CONSUMPTION BY MAJOR SOURCE, 1980–2000
(in millions of barrels per day of oil equivalent, or MBDOE and percent)

	1980 MBDOE	1980 %	1990 MBDOE	1990 %	2000 MBDOE	2000 %
Oil*	66	47	74	40	79	35
Natural Gas	27	19	37	20	43	19
Coal	36	26	50	27	63	28
Hydro	8	6	13	7	18	8
Nuclear	3	2	11	6	22	10
World Total	140	100	185	100	225	100

* The 1990 and 2000 data include 4 MBD and 9·MBD of synthetic oil respectively.

Source: Derived from data in Exxon, *World Energy Outlook (1980)*.

ESTIMATED WORLD ENERGY CONSUMPTION BY MAJOR SOURCE, 1980, 1990, and 2000.

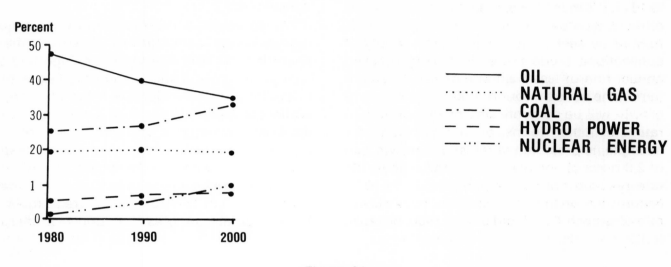

Figure 21.

old and new, that had not been explored, development of new sources of energy, and development of new technology. Given these considerations and given the ever increasing importance of government policies in the field of energy the outlook for energy supply seems to be as follows.

Total supply of energy is projected to grow at an annual rate of 2.4 percent between 1980 and 2000. This is rather lower than the growth rate of 2.9 percent reached during the period 1973 to 1979. It is much lower than the annual growth rate of 5.3 percent for the period 1965 to 1973. As can be seen from Table 38 the supply of conventional crude oil is expected to grow at an annual rate of less than one half of 1 percent during the period through 2000. This low rate of growth will be compensated by higher growth rates for other forms of energy. Coal, for example, is projected to register a growth rate of 2.8 percent per year, almost three times the rate reached during the period 1965 to 1973. Natural gas, on the other hand, will have a lower rate of growth than it did before 1980, because

most of the gas is produced in conjunction with the production of crude oil. By contrast to oil and natural gas, synthetic oil is projected to increase at a very high rate. Nuclear energy is also projected to increase at a rate that is much higher than the rate of growth in total energy supply. The differences in the growth rates of the various sources of energy will change the relative portions of these sources in the total supply of energy. These changes were discussed earlier in conjunction with the data in Table 36.

The projected supply of energy by major region varies considerably. The industrial countries will continue to be net importers of energy, and developing countries are expected to continue to be major exporters of energy. While the Centrally Planned Economies will increase their supply of energy, most of the increase will be directed toward domestic consumption with only a small part of the supply being available for export. It can be seen from the data in Table 39 and Figure 22 that for the industrial countries the supply of energy

Table 38.

ENERGY SUPPLY GROWTH RATE: PERCENT PER YEAR BY SOURCE, 1965–2000

	1965–1973	1973–1979	1979–2000
Oil	7.7	2.2	0.4
Synthetics and Very Heavy Oil	---	---	13.8
Natural Gas	7.3	3.6	2.6
Coal	1.0	2.4	2.8
Nuclear	27.8	20.9	10.0
Hydro	3.9	4.6	3.5
All Sources	5.3	2.9	2.4

Source: See source for Table 35.

Table 39.

WORLD ENERGY: SUPPLY AND DEMAND BY SOURCE AND MAJOR REGION, 1980–2000
(in millions of barrels per day of oil equivalent, or MBDOE)

	Oil	Coal	Natural Gas	Hydro	Nuclear	Other	Total Supply	Total Demand	Exports or Imports
Industrial Countries									
1980	14	17	14	5	2	—	52	80	28
1990	16	22	15	7	7	1	68	96	28
2000	16	30	17	8	14	2	87	109	22
Centrally Planned Economies									
1980	14	20	8	1	1	—	44	42	2
1990	15	24	13	2	3	—	58	55	2
2000	16	28	18	3	6	—	71	68	3
Developing Countries									
1980	38	2	3	1	—	—	44	18	26
1990	50	4	6	2	1	1	64	38	26
2000	42	6	13	4	2	3	70	48	22

Source: E. Ruttley, "World Energy Balances: Looking to 2020" in OPEC, *OPEC and Future Energy Markets* (London, Macmillan), 1980; Exxon, *World Energy Outlook* (1980); World Bank, *Energy in the Developing Countries* (1980); *World Development Report* (1980).

82

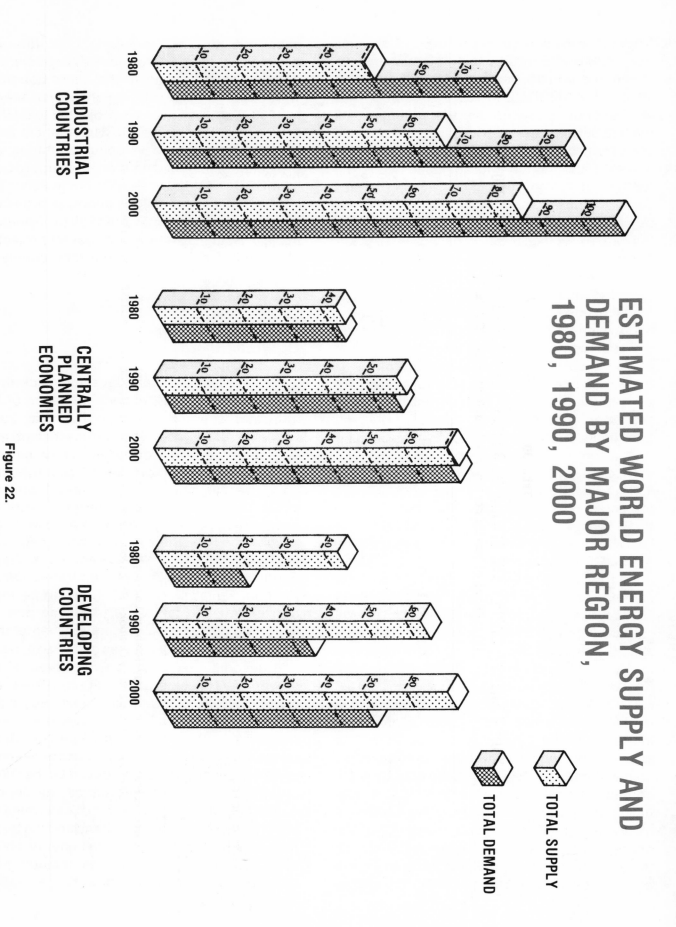

Figure 22.

ESTIMATED WORLD ENERGY SUPPLY AND
DEMAND BY MAJOR REGION,
1980, 1990, 2000

INDUSTRIAL
COUNTRIES

CENTRALLY
PLANNED
ECONOMIES

DEVELOPING
COUNTRIES

TOTAL SUPPLY

TOTAL DEMAND

from domestic sources is projected to increase from 52 MBDOE in 1980 to 87 MBDOE in 2000. Compared with a total demand of 52 MBDOE in 1980 and 87 MBDOE in 2000, these countries will continue to rely on imported energy to fill the gap between demand and supply. As can be seen from data in Table 39 the gap between domestic production and domestic needs will tend to diminish over time. In 1980 these countries were importing 28 MBDOE or 53 percent of their energy requirements. By 2000 they are expected to import only 22 MBDOE or 20 percent of their energy needs. One major development that is projected to contribute to the narrowing of the gap is the contribution of nuclear energy to total supply of energy. Nuclear energy contributed 2 MBDOE in 1980 or 4 percent of the energy supply of the industrial countries. By 2000 its contribution is projected to increase to 14 MBDOE, or 16 percent of the total supply of energy of this group of countries. The Centrally Planned Economies are projected to continue to meet their energy needs from their own domestic sources of supply, with a fraction of the supply finding its way to the world market. It is projected that these countries will increase their energy supply from 44 MBDOE to 71 MBDOE between 1980 and 2000, with some 2 to 3 MBDOE to be exported.

In terms of overall energy supply the share of oil tends to decline from 32 percent in 1980 to 23 percent in 2000, while that of natural gas will increase from 18 to 25 percent of total energy supply. Although the share of coal is projected to decline from 45 percent in 1980 to 39 percent in 2000, its production will increase from 20 MBDOE to 28 MBDOE during the same period. It should be noted that of the 27 MBDOE increase in the supply of energy about 18 MBDOE, or two thirds will be contributed by coal and natural gas. Developing countries will continue their status as net exporters of energy during the period 1980–2000. Oil will continue to be the largest contributing source in the total energy supply. It should be remembered, however, that oil will provide a declining share of the total energy supply. Also, net energy exports from the developing countries will decline in both absolute and relative terms. The data in Table 39 indicate that these countries were exporting about 59 percent of their energy in 1980. This proportion is projected to decline to 40 percent in 1990 and to 31 percent in 2000. In absolute terms developing countries are projected to increase their energy supply from 44 MBDOE in 1990 to 70 MBDOE in 2000. The share of oil in total energy supply is projected to decline from 84 percent in 1980 to 60 percent in 2000. Both natural gas and coal are projected to increase their share in the total supply of energy.

The Future Balance Of World Energy

It has been shown that developing countries will continue to meet the energy deficit of the industrial countries through the year 2000. There are 28 developing countries that are also energy exporting countries. They can be divided into two groups: the 13 countries that comprise the membership of OPEC and the other 15 countries. If one were to divide the energy exporting countries into Arab and non-Arab we find that there are 11 Arab oil exporting countries (Algeria, Iraq, Kuwait, Libya, Qatar, Saudi Arabia, United Arab Emirates, Oman, Bahrain, Egypt, and Syria). It would be more convenient if one were to divide the oil exporting countries according to whether they belonged to OPEC or not (see note, Part 1, pg. 23 and accompanying map). Only seven of the eleven Arab oil exporting countries are members of OPEC (Bahrain, Oman, Egypt and Syria are not members). These seven Arab countries account for about 95 percent of the oil produced in the Arab region and between 70 and 74 percent of the oil produced by the OPEC countries. World energy balance will depend on the outcome of the forces that determine the supply of energy on the one hand and the forces that determine the demand for energy on the other. The industrial countries, including the United States, will continue to be energy importers till the year 2000. The Centrally Planned Economies will continue to meet their growing needs for energy from domestic sources with a fraction of the supply directed to foreign markets, primarily markets of the industrial countries. The developing countries will remain the major net energy exporting group. Of the 118 countries that comprise this group only 28 will be net exporters of energy. These twenty-eight will be called upon to meet the energy needs of the other developing countries as well as those of the industrial countries.

Arab Oil and the Future Of World Energy

It was noted above that Arab oil constitutes about three fourths of OPEC oil. It is this oil that is projected to meet the future requirements of the world for oil over and above domestic production. An assessment of the future role of Arab and OPEC oil in terms of world energy should be preceded by three observations. First, several countries that are important oil exporters now will in the future lose their position of prominence either because of depletion of their oil reserves or because of the rise in domestic demand for energy, or both. This prospect is particularly true of countries like Algeria, Iran, and Venezuela. Second, it is projected that many developing countries will accelerate the development of their energy resources for domestic consumption. As a result, they will rely less on imported energy, though some dependence will persist. According to World Bank studies these countries are projected to increase their production of energy from 8.5 MBDOE in 1980 to 18.5 MBDOE in 1990. Their consumption of energy, however, is projected to increase from 14 MBDOE to 26 MBDOE during the same period. The projections indicate that these countries are expected to increase their imports from 5.6 MBDOE in 1980 to 7.5 MBDOE in 1990. Beyond 1990 or by the turn of the century they are projected to be moving close to energy self-

sufficiency as more domestic sources of energy are developed. The third consideration is the variation in the assessment of how much OPEC oil will be needed to fill the gap between supply and demand. Depending on the assumptions used, one finds considerable variations in the quantities needed. Some of these variations are shown in Table 40.

In all these studies one underlying assumption is that oil production of OPEC countries is residual or is a swing output. In other words, consumers come to them only after the consumers have used their own domestic energy supplies. The difficulty with this reasoning is that it assumes a passive attitude on the part of the oil producing countries in terms of their own conservation policies with respect to their natural wealth. This point was dealt with in Part III. Clearly, one of the most critical factors governing the availability of oil from OPEC countries is their own energy policies. These policies in turn will be determined by a multitude of factors including the countries' own needs for domestic consumption, which is projected to increase from 2.4 MBD in 1980 to almost 17 MBD in 2000.

Certain Arab members of OPEC are expected to be of critical importance to the supply of oil in world markets in the years to come. These are Saudi Arabia, Libya, Kuwait, Iraq, United Arab Emirates, and Qatar. These countries constitute a group designated by the World Bank as Capital Surplus Oil Exporters to signify the fact that they are the main oil exporting countries in the world. They are projected to export about 78 percent of all the oil to be exported in 1990 by all developing countries. It is possible to limit the list of capital surplus oil exporters to the Arab countries because Iran embarked on an oil conservation policy and because of its own increasing needs for oil. The case for excluding Iran was strengthened by the outbreak of war in 1980 between Iran and Iraq.

Given the present state of knowledge, technology, government policies and the cost of oil and non oil fuels, it is safe to say that Arab oil will continue to make a significant

Table 40.

ESTIMATES OF PROJECTED VOLUMES OF OPEC OIL

Organization Making the Projections	Year Projections Made	Required MBD	OPEC Production Year
Central Intelligence Agency	1977	47–51	1988
United States Congressional Research Service	1977	40–45	1985
International Energy Agency	1978	43	1985
Exxon	1978	40	1985
Common Market	1979	38	1985
United States Department of Energy	1979	28–36	1985
Exxon	1980	33	1990
Shell	1980	30	1984
Central Intelligence Agency	1979	30	1982
International Energy Agency	1980	31	1990

Source: Ian Seymour, *OPEC: Instrument of Change* (London, Macmillan, 1980).

contribution to meeting world needs for energy in the years to come. This is especially the case in view of the fact that means of transportation will continue to rely exclusively on oil as the source of energy.

Concluding Observations

From all current indications and studies there tends to be a general agreement that for the decades to come there does not seem to be any fear that there will be an energy crisis such as occurred following the October 1973 war. The mere growth in demand for energy does not justify the assumption that an energy shortage is in the making.

As to Arab oil there seems to be every indication that, for the balance of the twentieth century and beyond, it will continue to flow both to its traditional markets in the industrial countries and to developing countries.

Selected References

Abdel-Fadil, M. (ed.) *Papers on the Economics of Oil: A Producer's View*, Oxford, 1979.

Attiga, A.A., "The Economic Development of the Oil Exporting Countries" *Middle East Economic Survey*, October 12, 1981 (Supplement).

Blair, J. M. *The Control of Oil*, New York, 1976.

British Petroleum, *BP Statistical Review of the World Oil Industry, 1970 to 1979.*

al-Chalabi, F. J. *OPEC and the International Oil Industry: A Changing Structure*, London, 1980.

Exxon, *Middle East Oil*, 1980.

___, *World Energy Outlook*, 1980.

International Economic Report of the President, 1977, Washington.

International Monetary Fund, *Annual Report, 1973 to 1979.*

___, *International Financial Statistics Yearbook*, 1980.

___, *Government Financial Statistics Yearbook*, 1980.

___, *Direction of Trade Annual, 1971–1977 and 1980.*

Jacoby, N.H. *Multinational Oil*, New York, 1974.

Mabro, R. and Shihata, I.F.I., *The OPEC Aid Record*, Vienna, 1978.

Mabro, R. "Oil Revenue and the Cost of Social and Economic Development" in *Energy in the Arab World*, Kuwait, 1980.

McNamara, R.S., *Address to the Board of Governors, 1974.*

Mikdashi, Z. *OPEC States and Third World Solidarity*, Vienna, 1980.

Organization of Arab Petroleum Exporting Countries (OAPEC), *Annual Statistical Report, 1974 to 1978*, Kuwait.

___, *Secretary General Annual Report, 1974 to 1980*, (Arabic), Kuwait.

___, *Energy in the Arab World*, Kuwait, 1980, 3 volumes.

Organization for Economic Cooperation and Development, *Energy Policy: Problems and Objectives* (Paris, 1966).

Organization of Petroleum Exporting Countries (OPEC), *Annual Report 1975 to 1979*, Vienna.

OPEC, *Annual Statistical Bulletin, 1970 to 1978.*

___, *OPEC Member Country Profiles*, 1980.

___, *OPEC and Future Energy Markets*, London, 1980.

Ortiz, R. G. *Viewpoint*, Vienna, 1981.

Sampson, A. *The Seven Sisters*, New York, 1976.

Sayigh, Y.A., "The Social Cost of Oil Revenues" in *Energy in the Arab World*, vol. 1, Kuwait, 1980.

___, *The Economies of the Arab World: Development Since 1945*, New York, 1978.

Seymour, I. *OPEC: Instrument of Change*, London, 1980.

Shell, *The Oil Majors in 1979*, 1980.

___, *Energy in Profile*, 1980.

Shihata, I.F.I., *OPEC as a Donor Group*, Vienna, 1980.

Stocking, G. W. *Middle East Oil: A Study in Political and Economic Controversy*, 1970.

United Nations Conference on Trade and Development, *1979 Handbook of International Trade and Development Statistics*, New York.

___, *Financial Solidarity for Development: Efforts and Institutions of the members of OPEC 1973–1976 review*, New York, 1979.

U.S. Department of Energy, *Annual Report to Congress, 1979*, Washington.

U.S. Department of State, Bureau of Public Affairs, *Current Policy*, No. 132.

U.S. Senate Committee on Interior and Insular Affairs, *Presidential Energy Statements*, 1973.

U.S. Senate Committee on Finance, *World Developments and U.S. Oil Import Policies*, 1973.

U.S. Senate Committee on Foreign Relations, *Multinational Oil Corporations and U.S. Foreign Policy*, 1975.

___, The International Petroleum Cartel, *The Iranian Consortium and U.S. National Security*, 1974.

U.S. Senate Committee on Interior and Insular Affairs, *Geopolitics of Energy*, 1977.

U.S. Senate Committee on Energy and Natural Resources, *Access to Oil—the United States Relationships with Saudi Arabia and Iran*, 1977.

___, *Geopolitics of Oil*, 1970.

The World Bank, *Trends in Developing Countries*, 1973.

___, *World Bank Atlas*, 1977–1980.

___, *World Development Report*, 1980.

___, *World Tables*, 1980.

___, *Energy in the Developing Countries*, 1980.

PERIODICALS

OAPEC Bulletin
OPEC Bulletin
Middle East Economic Survey
Petroleum Intelligence Weekly
OPEC Review